● MONDEO THE STORY OF THE GLOBAL CAR

A WP&PC PUBLICATION

PUBLISHER:

WP&PC

- **World Publishing & Publicity Consultants SA**
  65, avenue de la Gare
  L-1611 Luxemburg
  Luxemburg

AUTHORS:
- **Arturo de Andrés**
- **Fulvio Cinti**
- **Vic Heylen**
- **Ray Hutton**
- **Jürgen Lewandowski**
- **Edouard Seidler**
- **Martin Wyler**

DESIGNED BY:
- **Rob Schouten, Rooduyn Vorm&Druk BV**

PRINTED BY:
- **Boom & Ruygrok Offset BV**

LITHOGRAPHY:
- **True Colours Nederland BV**

'MONDEO'/WP&PC
ISBN 2- 919983-00-8

Mondeo

The Story of the Global Car

Ford
Mondeo

The challenge was to create a global car, one which could be made in substantially the same form in Europe and America and would satisfy the needs of customers all over the world. For Ford, this was a new way of looking at things. The obstacles were formidable, the benefits irresistible.

TAKING A GLOBAL VIEW

● **Edouard Seidler**

The snow had gone and the trees were budding when the Board of Directors met in April 1990 on the top floor of Ford's World headquarters in Dearborn.

Sixteen men and one woman sat around the table. They included the five leading Ford managers, the two Ford heirs, Edsel II, the son of Henry II, and Billy, son of William Clay, and some of America's most prominent businessmen.

These people watch over the worldwide production of some 6 million vehicles per year - sometimes more, sometimes less. All knew that this meeting would be a long one, since its purpose was to study the most expensive and ambitious programme in Ford's history. Among other things, it was to call for an overhaul of several factories in the United States and in Europe, and would influence a large proportion of their world markets.

Harold ('Red') Poling was at the end of the table, in the chairman's seat which had once been that of Henry Ford II. Before he moved to the top at Dearborn, Poling - trained in finance - had presided over Ford of Europe. Chairing the meeting that day, he was more keen to run the discussion than to take part in it. The questions were left up to the Board; the pleading to Phil Benton and Allan Gilmour; the main answers to a few outside guests such as Lindsey Halstead, chairman of Ford of Europe, Alex Trotman, the man in charge of cars in the US, and John Oldfield, an Englishman, who was to head the programme.

Opening the discussion, Poling dwelt only on the main motivation of Ford's management in proposing their first 'global car': the harmonization of tastes, needs, and regulations that made such a programme possible; the potential savings that made it worthwhile; the respective competence of European and American Ford engineers that could make it happen; the necessity to make just one team responsible for it; and the arguments in favour of a 'European' team under John Oldfield.

One month earlier, Ford's top management had decided to move ahead with such a global programme, providing the Board gave it its green light. For three years before that, a number of studies had been conducted, primarily dealing with the requirements of American and European customers. The object was to kill two birds with one stone: to develop the replacement of Europe's Sierra - which was to be launched in spring 1993 at the Geneva auto show - while using the same car, with some styling modifications, to replace the Tempo and Topaz in the US the following year. They soon realized that the programme, which was to be known as CDW27 (CD for the car class and W for world), would produce a car up-market from Tempo and Topaz, two 'compacts' that remained popular with fleet buyers. It was decided, therefore, that the new car would not replace them, but would be an addition to the US range.

Allan Gilmour took over from Poling to explain and defend the programme. First he showed that the so-called 'world cars' of the Eighties - Ford's Escort and the GM Cavalier - were not really universal cars. They were quite different on either side of the Atlantic and had been developed by two independent teams which had hardly cooperated. The aim, this time, was to use just one team, which would call upon the specific knowledge and experience Ford had acquired in Europe and in the United States.

Lynn Halstead further clarified this point. He reminded the directors that before Ford of Europe was created in 1967, following suggestions by the late John Andrews, Ford had two car-making companies in Europe, one in Germany, and one in the United Kingdom, producing different cars and selling them to Ford's various subsidiaries. Though they were now harmonized by the existence of Ford of Europe, they still had two technical centres in Europe, at Merkenich in Germany and Dunton in England. Each had its specializations, but they cooperate - thanks to two daily Ford flights, to computer links and video conferences - in the development of the same cars for Europe as a whole: Fiesta, Escort, Orion, Sierra, Granada/Scorpio, plus derivatives of these models and light commercial vehicles.

'Why don't we do what we did in Europe but on a world scale?', Halstead pleaded. 'We can make use of all of Ford's potential. When one of our branches develops a small four-cylinder engine, it doesn't make sense for another subsidiary to develop its own. The same goes for transmissions, suspensions, ABS, catalytic converters, and, generally speaking, all components, whether we make them or we buy them from suppliers.'

Benton nailed it in : 'When you deal with the same type of clients on both sides of the Atlantic, why make two cars when one would be enough?'

Gilmour went on to list the changes which made a global car possible:

- the tastes of clients in different continents have moved closer, to the point where, as Benton put it, 'there are no differences between them any more, just nuances';
- safety and exhaust emissions rules are ever-more harmonized;
- on both sides of the Atlantic, buyers tend to pick cars that are lighter and use less fuel - economy is becoming a worldwide issue;
- the problems of, and demands for recycling used cars are the same in Europe and in the US;
- computers, satellite TV, and jets have reduced the size of the world and are making technical communication easier;
- the development cost of a new car, especially a mass-market product (the plan was to produce 650,000 CDW27's per year, 400,000 in Europe and 250,000 in America) can be reduced greatly if it produces several models;
- it is a similar story for purchasing new tooling and making economies of scale by grouping the development and purchase of components at the same suppliers.

All in all, Gilmour insisted, it had become easier to conceive a global car. Everything was in favour of such a programme, though it was not without difficulty, complexity, and risk. It was obvious, he continued, that Ford would find it more difficult than the Japanese, for they developed their cars in a central location and their overseas technical centres simply adapted existing cars to local requirements, sometimes selecting different components for them. CDW27 would go way beyond

The tastes of clients in different continents have moved closer. Interior design reflects the car's global aspirations.

Safety rules are becoming harmonized throughout the world. Ford is one step ahead by fitting a driver's side airbag as standard throughout the range.

that, but its potential made it worthwhile.

'There is quite a difference', Halstead suggested, 'between global sales and global manufacturing...'

Without even going into detail - that would be Oldfield's responsibilty - it appeared that there were benefits for both Europe and the US. Ford of Europe had developed the Zeta, a modern 16-valve four-cylinder engine; Europe is also good at developing cars that handle well. The US, on the other hand, must be leaders in terms of V6 engines, automatic transmissions, power steering, and response to emissions regulations. All of that seemed to form a good base for fruitful cooperation.

It was also clear that Ford of Europe's engineering team was best qualified to handle the CDW27 programme. 'Europeans', Benton said, 'are used to cars of this type. They know, better maybe than we do, what their role is in society. It is logical to entrust them with such a programme.'

So, the Europeans would be in charge. After a discussion which went on for one and a half hours, the Board accepted the CDW27 programme.

In the months that followed, Oldfield would often meet his superiors in Dearborn to report on how the programme was progressing. He did, of course, have the competition screened closely.

'Which', they asked him, 'are the best cars from our competitors?'

Oldfield produced a list.

'Be careful', they then told him. 'We don't want to be at the level of the best in 1990 or 1991. We must be the leader much later; that is when our own cars are available. And we must be able to answer those who have reacted in the meantime'.

'Of course', Oldfield replied.

'And don't forget our goal. While a client satisfaction rating of 80 is good, we are aiming at 95 with this car'.

As a group, Ford's top management travels to Europe three times a year. Louis Ross, when he was executive vice-president in charge of overseas divisions, made three more trips yearly. In between, there are HQ meetings at Dearborn.

At each meeting, for the next two years, they reviewed the global programme. There was no need to send a large number of specialists overseas; they used video conferences instead of planes.

There have never been more than 35 Americans at any one

time working on this project within Ford of Europe. Ford HQ delegated a few purchasing people, a couple of finance men, and a number of process and manufacturing engineers. They worked, in a new building set up next to the Genk factory in Belgium, with their European colleagues, representatives of the suppliers, and research and development engineers brought in from England and Germany. Genk also had a pilot plant, where these men - working side-by-side long before production of the new car was to start - could prepare for the future, while testing their methods and skills.

After European production start-up, a number of these specialists would travel to Kansas City to prepare for the launch of the US version of the CDW27.

The replacement for the Sierra will have a new name in Europe, Mondeo, which well illustrates its worldly traits. Its American names (that of the future Ford and of its Mercury cousin) have yet to be chosen.

Mondeo will be more conservative in style than the Sierra was for its time - and which initially created some sales problems in the UK. Ford's various styling studios, including the one in California, all contributed their concepts for this new first global car.

Of course, no cumbersome and heavy sheet metal will cross the Atlantic. Stamping will be done on both sides of the water, using the identical tools.

The US will have only four-door versions, while the European car to be introduced at the Geneva Show has three different body variants : four door saloon, five door hatchback, and estate car.

Both the American and the European cars will use the same platforms, the same bodies (with minor styling differences at front and rear), the same suspension with different settings, and the same engines and transmissions.

The 2.5 litre V6 engine, not a major factor in Europe, will be imported from the USA, made at Ford's factory in Cleveland,

A new badge - the name Mondeo illustrates the car's worldly traits.

Ohio. That will be a year after the European version is launched with four-cylinder engines manufactured in Bridgend in South Wales. Ford Mexico will send it's Zeta engines to America to power Mondeo's Ford and Mercury sisters. Ford of Europe alone will use a diesel engine manufactured in its factory at Dagenham, England.

The manual transmissions will all come from Europe, the automatic ones (four-speed with electronic controls) from the US.

For what will be bought outside, Ford gave preference to international suppliers, capable of producing the same components on both sides of the Atlantic. Quite a few components will, however, be sourced in one place only and must cross the ocean: 47 European suppliers will provide 147 different parts and components to the Kansas City factory, while 20 will deliver their goods from America to the Genk factory in Belgium, the only European plant to assemble Mondeo.

They might have gone even further had Ford not wanted to maintain a high proportion of 'local content' on each side of the Atlantic.

Surely it has been necessary to accept a number of compromises to achieve such a global programme? Ford managers claim not. Poling, for one, says there have only been 'compromises on detail'.

The car went through thorough 'clinics' everywhere. There were three in the US: in California, in the East, and in the Mid-west, and several others in Europe (Germany, England, France, Italy, Belgium, and Spain). The conclusions from all were in favour of the programme.

Some apparently small details assume great importance at these research sessions. For example, cup holders were considered an absolute must in the US, a request with which Ford complied and from which European customers will also benefit. The radio buttons were criticized as too close to one another for a driver with gloves in winter. So, they were sited further apart!

Fortunately, Americans and Europeans agreed on the essentials, demonstrating a harmony of tastes and needs - as Ford management had predicted.

What next? Poling admits that at least one more global car is being developed, which could go one step further and involve Mazda of Japan, in which Ford has a 25 per cent shareholding. 'It was impossible', Gilmour says, 'to start out coordinating Ford

US, Ford of Europe, and Mazda. In future, however, we might well develop a truly universal car, which could be produced at the same time in the US, in Europe, and in Japan'.

For the time being, Ford is considering exporting Mondeos to Asia, where Allan Gilmour recently assessed their potential and American CDW27's to the Gulf States.

Ford's key managers admit that the 'concept-to-customer' time - that needed to develop a new model - and the life cycle of any new car will be greatly reduced in the future. There will also be more revisions and modifications during a model's life-time. (The Sierra lasted over ten years, with one major 'facelift'. At its peak, production exceeded 350,000 units a year.)

In the following pages, some of Europe's best automotive writers analyse the main aspects of the new Mondeo. We hope that the buyers of this global car will enjoy driving it as much as we enjoyed getting to know it through our conversations with some of its 'fathers'. ●

CDW 27
STRICTLY
CONFIDENTIAL

It is one thing to decide to do a global car, another to make it happen. Once the CDW27 programme team was in place, they began drawing up targets for the car. It would face strong competition from European, Japanese and US manufacturers. It had to be better. But how much better? How far could they stretch?

THE PLAN EVOLVES

● **Ray Hutton**

John Oldfield remembered Ford's last attempt at a global car. They had tried it with the Escort in 1981, when he was in charge of product planning in Ford of Europe. The cars turned out well enough but the company did not achieve any benefit from using the same basic design on both sides of the Atlantic. By the time they came to the market, the only common parts between those first European and US front-wheel drive Escorts were a water pump seal and the Ford oval on the grille. That was, perhaps, inevitable, as there had been two separate engineering teams. Oldfield realized that for a global programme to be truly effective, there had to be only one team of engineers.

He had this in mind when he was called to Phil Benton's office one morning in November 1985. A Ford veteran of 27 years service, Oldfield had recently been promoted to executive director of engineering staff at the company's World Headquarters. He had moved from his home in Rayleigh, Essex, and was just settling into a comfortable house in Bloomfield Hills, a smart suburb of Detroit.

Benton, then head of the Ford's automotive operations, asked Oldfield to undertake a special project: 'We want you to take another look at the benefits of world-wide model programmes. Do a detailed study.'

In the Ford world, a global programme could be two-way or three-way, involving either all or some combination of Ford of

America, Ford of Europe, and Mazda Motor Corporation. Oldfield's study concluded that involving all three would be ideal but at that time lines of communication and working practices were not well enough developed to cope with a three-way programme. Of the two-way links, a joint effort between North America and Europe was the most promising.

This pre-supposed that the market needs were similar. If not, the global car would end up as an unsatisfactory compromise.

As it happened, the need to replace the European Sierra and the US Ford Tempo and Mercury Topaz came together. These cars were about the same size, similar in accommodation, and sold in comparable volumes. Might they, or rather a single

car to succeed them, be a candidate for a global programme?

Oldfield and his team began to study the two markets and the demographics of buyers of cars in what Ford call the CD class. The North American sales organisation wanted a car up-market from the Tempo, in the upper CD segment, which was near to a European specification. The deeper they looked, the closer the requirements seemed. The case for a 'CDW' project -a CD car for the World - was strong enough to recommend to top management.

In June 1986, the OCE (Office of Chief Executive) - the triumvirate of chairman, president and chief financial officer agreed to Oldfield's proposal. CDW27 (the number has no significance) became a global programme, embracing Europe

Ford's future starts to take shape - an early rendering for CDW27 shows the starting point for a global style.

and North America.

The CD class had a greater significance in Europe, where there was more experience of cars of this type, so Ford of Europe would lead. But who would head the programme? Who better than John Oldfield, the man who had worked out the strategy...

This meant a move back to England for Oldfield. But not yet. He started to put together an engineering team, drawn from Europe and North America, but would head it from Dearborn for most of another year, transferring to Europe when the project moved into the next stage, in March 1987.

Ford of Europe were already engaged in some pre-programme work for a new Sierra, slated for launch in 1993,

Styling of a new project starts, as it always has, with pen-on-paper sketches.

The latest computer techniques in the design studio allow shapes to be built up and reviewed on screen.

while in America there had been plans for a replacement Tempo.

Looking back, these cars were very close to what CDW27 became. Of similar size, both were front-wheel drive with transverse engines. Tempo had been front-wheel drive since 1985 but Ford of Europe regarded the CD class as the cross-over point between front and rear-drive and had continued to debate the advantages of both layouts.

Oldfield recalls that this time they quickly concluded that CDW27 should be front-drive. The car's 'packaging' could then be better and there would be fuel economy benefits, thanks to greater drive-line efficiency and lower weight. All manufacturers in the USA are concerned about CAFE (Corporate Average Fuel Economy) and it was important there that the new model contributed to lowering Ford's overall fuel consumption figure.

It was clear from the existing products in the upper CD class, many of which were Japanese, that Ford's contender for the mid-1990s would have to have a level of technical sophistication way beyond the Sierra, the GM Cavalier/Vectra, or the Volkswagen Passat. Of course, the competition would not be standing still, and the engineering chiefs in Russelsheim and Wolfsburg were facing the same challenge.

More sophisticated, but how much more? How far did they need to go? Before the CDW27 programme could start in earnest they had to have some agreed parameters. They had to be sure that the engineers in Europe and North America were speaking the same language, valued things in the same way.

During 1987 existing competitors to Sierra and Tempo/Topaz were brought together and assessed by teams in Europe and the US. Gradually attention focussed on a car that was sold in substantially the same form on both sides of the Atlantic and manifestly satisfied both US and European customers: the Honda Accord.

The Accord, built at Honda's plant in Ohio as well as Japan, was a symbol of how things had changed in the American market. It was to take the place of the Ford Escort as the best-selling car in the US. But it was convenience rather than any corporate sensitivity that was to make the 1989-model Accord the benchmark for CDW27.

For all dynamic characteristics the Accord was rated 100 per cent. Though it was accepted as the best car overall, this did not mean the Honda was superior in every respect, nor that Ford's objective was simply to build a better Accord. It was just a starting point, the best way of establishing common ground.

Within Ford, drive assessments use a 10 point scale, the higher the number, the better the car. Existing Ford models and a range of competitors were given extensive track testing to establish scores in the key dynamic parameters: ride, steering, handling, power-train NVH (Noise Vibration and Harshness), body/chassis NVH, transmission shift quality, performance feel, driveability, brake feel.

Whatever the Accord scored became 100 per cent. It might, for example, have been given 7 points for ride comfort. But another car, the Toyota Camry, had the best initial rating for ride. It scored, say, 7.5, half a point higher, so that was 105 per cent. The judgement then was where should CDW27 be? Was equal to the best in class good enough? Or should they aim higher, and if so, by how much? 5 per cent? 10 per cent?

Ford engineers talk about the higher targets as putting 'stretch' into the programme. Whether that can be justified depends on the rather less precise business of trend analysis. This means predicting where the competition will be five years ahead and setting targets above them.

Since a greater stretch usually means a higher level of technology, and thus more expenditure, the engineers have to put a persuasive case to the company's management.

John Oldfield recalls some spirited discussions with Harold Poling about the value of some of the car's features. But he takes pride in the prevailing corporate attitude: 'Top management said that we are in a business where the best product wins. We must put in whatever is necessary to make the car very competitive and then find ways of making it profitable.'

Meanwhile, in shaping CDW27, Ford's designers were

striving to meet another challenge. Donald Petersen, Poling's predecessor as Ford chairman, had asked the company's design studios worldwide to identify the next design trend, to anticipate the style of cars that would be popular in the second half of the 1990s.

They concluded that 'cab forward' was the way to go. This meant moving the car's cabin further towards the front of the car, a low and short bonnet, and minimum 'overhang' in front of the wheels. Apart from achieving a distinctive look, this also promised more room inside the car - though less space for the front-wheel drive power train under the bonnet.

It is notoriously difficult to get customers to accept futuristic ideas. And therein lies a problem for the CDW27

Market research 'clinics' play an essential role in establishing the acceptability of new concepts - here CDW27 concepts are scored against existing competitors.

exterior design team, based at Merkernich. Their target was achieve best-in-class ratings at market research 'beauty clinics' where members of the public are shown new, unbadged designs alongside familiar models.

In 1988, the car that was winning the beauty clinics, in Europe and the US, was the Honda Accord. That was a straight up and down 'three box' saloon, far from the cab forward ideal. The softer, more curved shape of the CDW27 proposals did not score well.

As time went on, there was a convergence of opinion. Ford's design proposals were modified - made somewhat less radical - and the clinic results swung towards the new styles. The mechanical heart of CDW27 was being tested in rather less glamorous surroundings. What the engineering fraternity call 'concept cars' are a long way from the shiny models of the design studios and motor shows. Early in 1988 the engine, transmission, suspension and steering being developed for CDW27 were installed under normal-looking Sierra bodywork. Eleven such cars were built and would be used to establish

Shaping up - CDW27 full-size clay model is formed in the design studio.

MONDEO DEBATED ON TV

John Oldfield is proud of Ford's video conferencing facility. Just outside his office at Dunton is a fully-equipped television studio. They don't make TV programmes there; they have meetings. Important meetings, with colleagues using similar facilities hundreds or thousands of miles away.

The studio is arranged somewhat like a boardroom but on one side of the table, instead of seats there is a large TV screen. Cameras and microphones are arranged so that all participants can talk normally, and show drawings and flip- charts, to those on site and far away. The meeting can be chaired just as if everyone is in the same room.

Video conferencing was started to improve communication between Ford's European technical centres in Dunton, England and Merkenich, Germany. When CDW27 was established as a global programme it was clear that a transatlantic video link to Dearborn was also needed. Now, even senior management meetings, like those of the powerful technical affairs committee, can be conducted by TV. Dunton's video conferencing studios are booked for up to 16 hours a day.

'Without video conferencing, the amount of travelling involved and the time differences would make a project like CDW27 damn near impossible', says Oldfield.

whether targets could be met and therefore turned into objectives for the CDW27 programme.

In October 1988, John Oldfield was made executive director of programme offices, on his way to being appointed vice-president in charge of all product programmes, vehicle engineering, and design in August the following year. When Oldfield took charge of all product programmes, David Price took over as director of the CDW27 project.

David Price had been director of powertrain programmes and as such was already involved in CDW27. He had been instrumental in up-rating Ford's new four-cylinder engine from the Gamma project, with three valves per cylinder, to the 16-valve Zeta, that was to be the mainstay for the Sierra's successor

as well as power unit for higher specification Escorts and Fiestas.

CDW27 programme office, the co-ordinating hub for the car's development, consisted of around 50 people, a dozen of them based in Dearborn to ease co-ordination with North American manufacturing, as well as sales and marketing. Throughout Ford of Europe engineering staff, the project was organized into a series of working groups, bringing together skills and knowledge to concentrate on each element of the car. Often these groups also included representatives of component suppliers. Some were essentially North American, for it was decided early on in the programme that some parts and systems - V6 engine, automatic transmission, power steering - would be

'Cab forward' becomes reality - a prototype comes close to the definitive CDW27. Gradually, the clinics, conservative at first, came round to approving the new style.

sourced from the United States.

The chairmen of the working groups - some 25 in all - reported regularly to a programme control meeting (PCM). From this, Price was able to glean the information he required to report progress to the special products committee, chaired by Lindsey Halstead. As chairman of Ford of Europe, Halstead had ultimate responsibility for CDW27's development.

By March 1989 Halstead was ready to set out the programme objectives to the technical affairs committee, made up of the OCE. Price arranged a drive appraisal in the secrecy of the Lommel proving ground in Belgium. Don Petersen, Red Poling, Phil Benton, and Allan Gilmour tested the concept cars, assessing all their CDW27 features. This was no mere rubber-stamping but a constructive debate. Price remembers that most of the discussion was about stretch and that the company's top executives commented that in some areas the engineers had not gone far enough while other targets were too ambitious and unrealistic.

Taking heed of the OCE's views, Price proceeded to draw up the details - including precise cost estimates - for what is known within Ford as prime programme approval. Price calls this 'the tough one'. It is regarded as the key checkpoint in a new car's development, the time when alternatives are dismissed, all the major parameters laid down, and its appearance was agreed. For CDW27, it was in June 1989.

It was still possible to change or improve the specification; indeed, it was vital to retain this flexibility, for every new car launch by a competitor could move the goal-posts from the points established at the original competitor assessments. In most respects, the CDW27 team had stretched far enough with their objectives to keep them at the top of the class, but the arrival of the Nissan Primera in 1990 did demand a re-think in the area of powertrain NVH, or refinement. Ford's objective was revised.

Price refers to objectives with a capital 'O', as hewn-in-stone, required to be achieved in full, with no excuses. By March 1990 all the objectives for the CDW27 programme were brought together in what is known as the 'blue letter'. This is actually a substantial book and it is not blue but white. It is the new car's bible.

As programme director, it was Price's job to compile and present the blue letter. The objectives fall into 18 different categories and apart from including dynamic and performance standards, they include timing and quality, safety and exhaust emissions, serviceability, insurance ratings, build complexity and volumes, costs and profitability.

Presentation of the blue letter to the OCE constitutes formal programme approval. A month later it would be put to the full board of directors for ratification.

The first prototypes, correct mechanically but still not fully representative in appearance, were built in September 1989. They were followed by accurate evaluation prototypes early in 1991. Because of the complexity of the programme, CDW27 had more prototypes than ever before - 400 would be built - continuing through to engineering sign-off in February 1992. Of these, 240 were European models and 160 the North American version.

From then on there was always the risk of an information leak, of a chance snap taken of a disguised prototype on test or in transit. Inevitably, such pictures soon appeared in the world's motoring press. In the absence of official comment, there was some confusion about which model was which. This was not surprising really, for while Europe was to have a five-door hatchback, four-door saloon and five-door estate car, only the saloon was planned for North America - and there will be two distinct versions of that, to be badged Ford and Mercury.

Job 1 - the point when the first saleable production car comes off the line - at Genk was on 23 November 1992. Seven weeks earlier, in Paris, at a dinner for the world's automotive press in a restaurant in the Bois de Boulogne, Allan Gilmour revealed the jealously guarded secret of its name, as the first images of Mondeo flashed across huge video screens.

At that point, in Europe at least, the development of CDW27 was complete. The countdown was on to bring Mondeo to the market in March 1993. ●

The chairman had given his designers a challenge: anticipate the trends, find the shape and style of the future. They agreed that 'cab forward' was the way. The problem then was to reconcile the new idea with the more traditional demands of today's customers - on both sides of the Atlantic.

SHAPING THE FUTURE

● Fulvio Cinti

The shape and style of CDW27 was 'frozen' in June 1989. It was two years and nine months from the outset - not long, considering the importance and scope of the project which, as a 'global car', would have to satisfy the demands of customers from different nations with differing outlooks and expectations.

Manfred Lampe, design programme manager, describes CDW27 as the biggest styling or design project ever undertaken by the Ford Motor Company. Lampe is German and he built up his professional experience in Ford's US and European studios, including four years with Ghia in Turin - the Ford subsidiary considered in design circles as a merit-worthy 'ideas pool'. Before being appointed to the CDW27 programme, he was on the design staff at Dearborn, working for Jack Telnack - Ford's design vice-president.

Lampe held the reins of the programme throughout the entire evolutionary process, from autumn 1986 until February 1992. He says: 'Our philosophy is to put our money on design, supporting and encouraging creativity within our staff, balancing market needs against advanced, leading-edge design. The question is - how to achieve that?'

For CDW27, the design operation started with four concept proposals being commissioned from the Carrozzeria Ghia in Turin, Ford's California Concept Center in San Jose, and the design studios in Dearborn and Cologne-Merkenich.

Just five months passed from initial, freewheeling sketches to concept models and the first review meeting in March 1987. The gathering of the European and American design teams, which was continued at the company's worldwide design conference in Cologne in April, had a precise purpose: to read and interpret the strengths and weaknesses of the four studies and establish which of the models would form the starting block for the 'global car'.

At these meetings, details of each proposal were analysed, their proportions and stylistic solutions weighed up, the most satisfying aesthetic elements identified, and references to the Ford image pin-pointed. In the end the 'target car' elected was that proposed by Cologne - actually, a product of several Ford of Europe studios, including Dunton.

But, unlike similar confrontations of models conceived by the company's design centres with others from external collaborators, at this Cologne meeting there were no winners or losers. The ideas expressed by the four cars fused together to compose a reference scenario for CDW27's design philosophy.

It is illuminating to discover how those ideas were assessed. The Dearborn car was seen as 'up-market', with its top-of-the-range feel expressed by a continuous belt-line and body side-mouldings, high foreline and generous brightwork (lights, mirrors, bumpers, handles, chromed window frames).

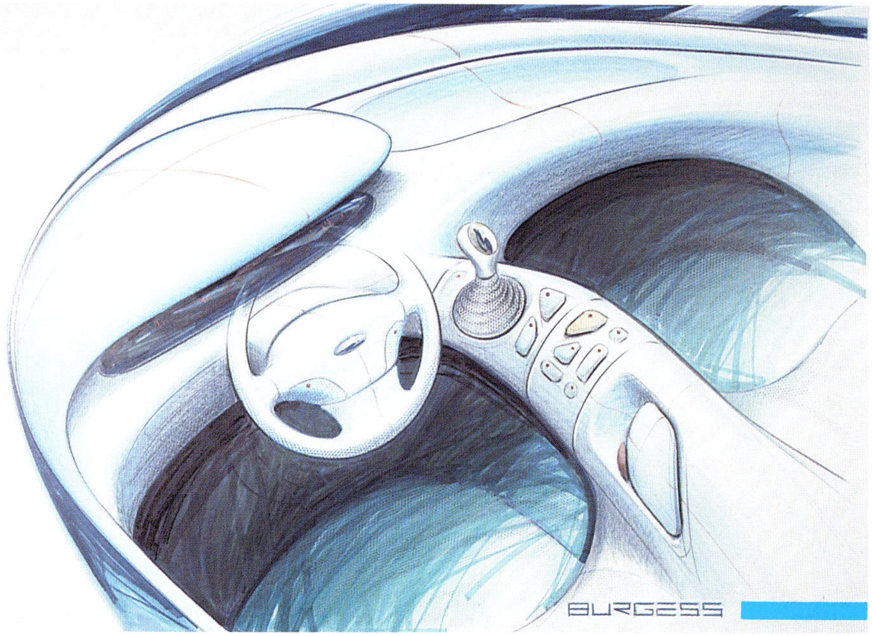

Free thinking - Ideas for a new car are first expressed in hand sketches, like this futuristic idea for the interior.

The large, high tail-lamps proved controversial. Longest of the four, the car was nevertheless perceived as too small for the US market.

The California car, conceived and built by the California Concept Center - a studio, under the direction of Richard Hutting, which proposes three or four concept models for Ford each year - expressed a more advanced form, with a strong sports image deriving from its fluid and continuous belt-line, glass 'greenhouse', and integrated bumpers. It was not, however, seen as a family car. The headlamp shape was controversial and the rear deck was criticized at the time as being 'soft and over-inflated'.

'Inoffensive' was the prevailing comment on the Turin proposal, conceived and constructed by Carrozzeria Ghia under the direction of Filippo Sapino. If that sounds like faint praise, this car also gained approval for its European 'feel', and, of the four proposals, it was the one that 'filled out the cube' most effectively, meaning the best perception of proportions. There

Ford design studios include the latest computer systems which enable renderings to be created and modified on screen.

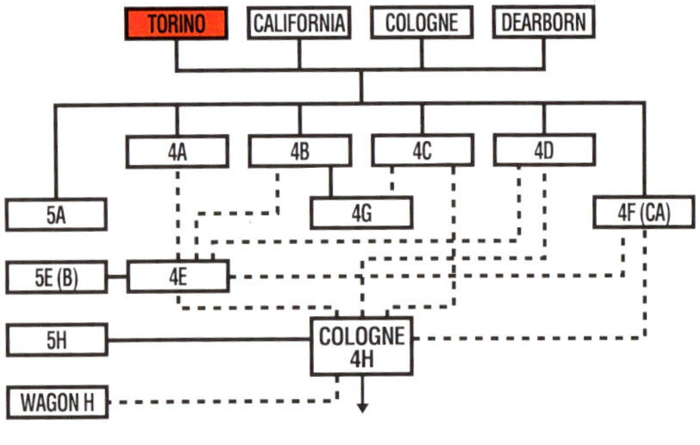

Initial concepts for CDW27 are depicted on the next eight pages. As the chart, left, shows, there were four proposals. The photographs to the right are of that made by the Ghia studio in Turin.

The drawings on these pages are from the early design phase but do not necessarily relate to the accompanying concept models.

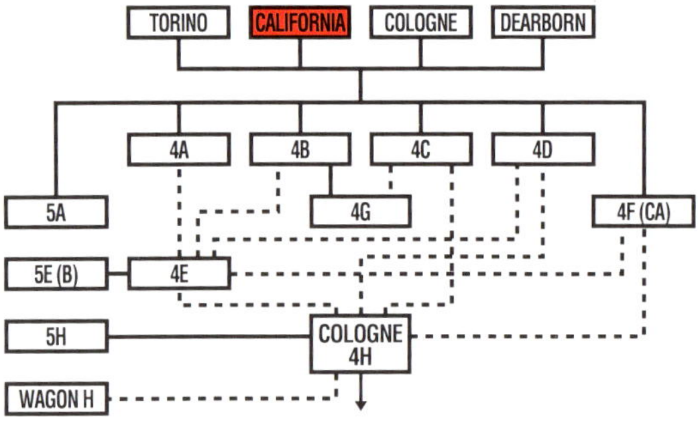

TORINO
CALIFORNIA
COLOGNE
DEARBORN
4A
4B
4C
4D
5A
4G
4F (CA)
5E (B)
4E
5H
COLOGNE 4H
WAGON H

West Coast influence - the second of the initial studies was from Ford's Concept Center in California.

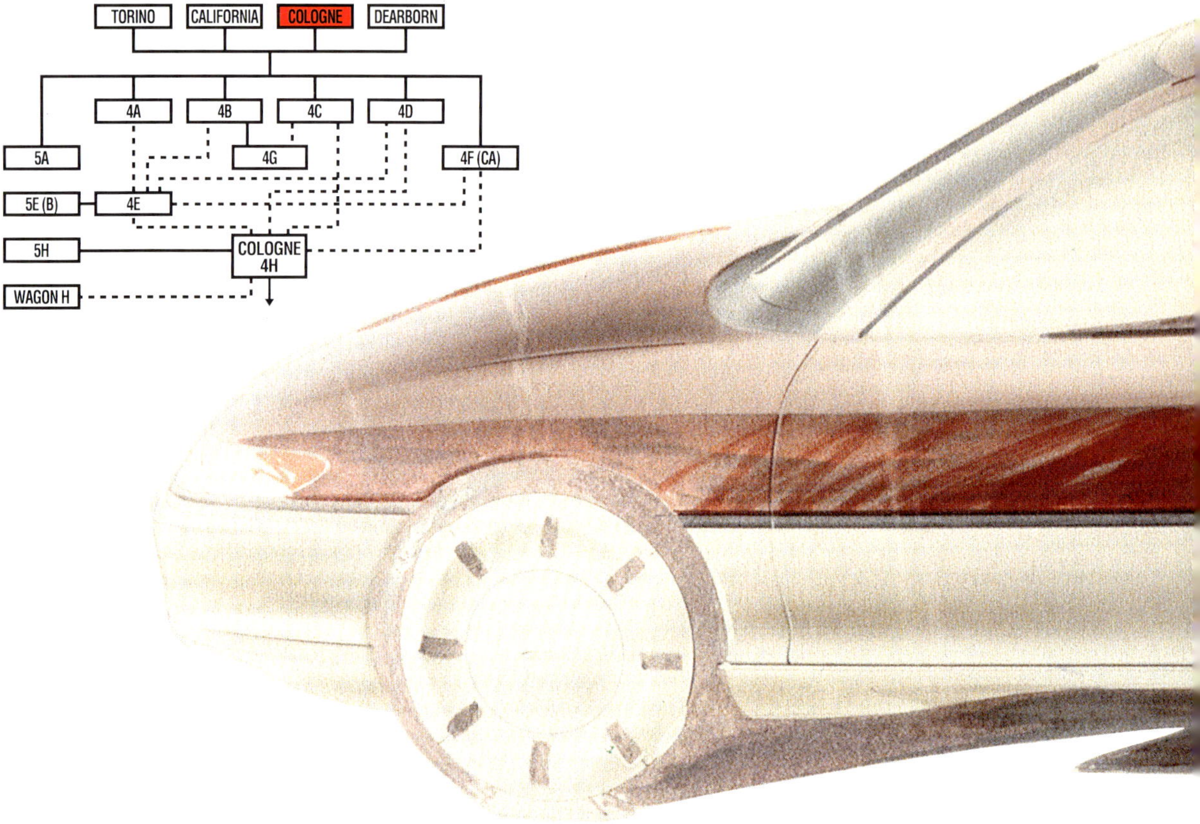
TORINO
CALIFORNIA
COLOGNE
DEARBORN
4A
4B
4C
4D
5A
4G
4F (CA)
5E (B)
4E
5H
COLOGNE
4H
WAGON H

The Cologne proposal, which also had input from the Dunton studio in England, was selected as the basis to proceed to the next design stage but would absorb ideas from the three alternatives.

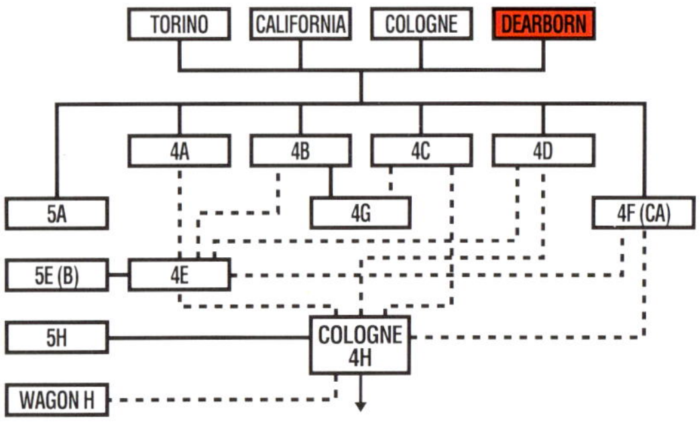
TORINO
CALIFORNIA
COLOGNE
DEARBORN
4A
4B
4C
4D
5A
4G
4F (CA)
5E (B)
4E
5H
COLOGNE 4H
WAGON H

Fourth of the initial design proposals was this more formal style from the studios of Ford headquarters in Dearborn, USA.

was some criticism of details, such as the body-colour B pillar and over-complicated bumpers.

The Cologne study was agreed to have a sporty image, with its tautness and large wheel openings giving it a light feel. Though it was the chosen starting design, it did not escape criticism. The rear end was too dominant, seeming to shorten the car, and the step in the belt-line and its heavy bumpers were considered ill-judged. At the Cologne conference, concepts for the car's interior were also examined. Some initial work had been done on the general 'package' and ergonomics, with three sit-in bucks constructed and several partially-clad foam-board instrument panels made to assess the bulk and form of the facia. At that time six panels - three for North America and three European -were being developed in clay. One key requirement was to visually minimize the apparent depth of the panel which was a result of the low, far forward cowl of the preferred exterior design.

The different concepts would later be presented for market

Reviewing an instrument panel design - modelled in full-size - so that ergonomic as well as aesthetic aspects can be assessed.

research in Germany and the UK. For those sessions, they would build sit-in bucks with windscreens and roof-forms cut from CDW27 exterior models, so that respondents could evaluate the driver and passenger environment more realistically.

In the spring of 1987, when development of the CDW27 programme began, Ford of Europe's design director was an American, Andy Jacobson. He had taken over some time before from Uwe Bahnsen, the man who had, for a long period, presided over the design of all Ford's European products and was considered the 'father' of the Sierra's futuristic form. Jacobson commanded the studios in Merkenich, led by John Doughty, and Dunton, under Jim Arnold.

Developing the exterior form of CDW27 thus became the task of the Cologne studio, while it was up to Dunton to develop interior themes. In the course of the evolutionary process the Dunton studio would also put its name to an exterior model conceived by their advanced design team.

The Cologne team started to develop models, calling them 4 and 5 respectively for the saloon and hatchback versions. The original Cologne study had established the basic exterior form for CDW27's programme description book. Lampe describes the next phase of the development as 'architectural', settling the overall proportions - the length of the bonnet, the position of the greenhouse, the slope of the windscreen, the proportion of

FORM FOLLOWS FUNCTION

Where is the most convenient position for the lights switch? Should electric windows be operated from the door or the central console? What sequence of windscreen wiper action do drivers prefer?

Ford's designers and engineers had a bank of such detail information available during the development of CDW27. It was obtained from thousands of ordinary motorists in Europe and the USA in a continuing research programme that seeks to establish basic customer preferences about cars and their use.

Ford calls this Expanded Image Research. As many as 15 vehicles of several makes and kinds are collected in one place and members of the public are invited to examine them and give their views.

Questions cover 700 parameters, everything from seat shape and switch action to visibility and visual appeal.

'This is a rolling index of customer awareness. It tells us what people like and don't like', says CDW27 programme director David Price. 'Expanded Image Research is a working tool that we use as a check throughout our development programmes. It played an important part in the CDW27 project'.

width and height to length.

The interior package progressed in parallel. Both aspects had to consider the latest legal requirements for all world markets, side-impact measures, bumper structure, and interior protection.

Only after all these matters had been settled could the 'styling phase' begin in earnest. Helmut Schrader who has since taken over from Doughty as head of the Cologne studio, affirms that establishing the car's proportions was a key aspect of development: 'We can only achieve fundamental newness in the marketplace if we radically change the proportions. If you look at the dimensions and compare them with the Sierra, the car certainly looks a lot different. It's much wider, much lower, with the cowl and bonnet height reduced. The wheelbase is longer by about 10 cm, although the two cars are more-or-less the same length overall.'

For programme manager Manfred Lampe it was a hard struggle to get these new 'cab forward' proportions - and the new problems and challenges they presented - accepted by the engineering and manufacturing departments. Lowering and shifting forward the cowl was combined with a new power-train arrangement - transverse-engined and front-wheel drive, rather than rear-drive like the Sierra. 'The emphasis on proportions forced the engineers to find new ways of packaging all the components very tightly together whilst still achieving the quality targets. In effect, the design also influenced the mechanical layout of the vehicle', says Schrader.

If the architectural aspect was vital in determining the optimum use of space for the mechanics and the cabin, the exterior styling was essential for attracting the customer. Lampe and Schrader were determined that it should be better than the opposition - but without absorbing external influences: 'We did not try to be Japanese. We did not try to be American. We tried to develop our own style, regardless of origin or heritage.'

Any influence there has been is surely German. Schrader rejoins, 'I wouldn't go as far as saying that CDW27 has a German style. Our cars set a certain trend at the time with the Sierra and the Scorpio and I think CDW27 is a natural progression from that. It's true to say that the Cologne model was the one that was picked for further development, but it's just as true that it was the influence from the other models that changed it to what it eventually became.

'But if you regard German styling as being conscious of

creating a car that has stance and good proportions and sits well on the road - well, yes, I think it's German design. German design puts functionality on top, and our car is functional. But it also has a bit more style and a bit more flair, perhaps some French or Italian influence. I think we have a very nice blend: Germanic attributes but a Latin character.'

The creative input from the Merkenich and Dunton studios took account of all the information culled from the various market research clinics, but this did not inhibit their freedom to envisage more advanced solutions. From their drawings and small-scale models derive the four preliminary full-scale models, produced and reviewed over the period from late July 1987 to early June 1988. Three were built at Merkenich and identified with the codes 4A, 4B, 4C while the fourth was prepared at Dunton and coded 4D. Each model was first made in clay then later in glass-fibre.

For those unfamiliar with the automotive design process and its exacting work methods and selection procedures, a period of 12 months to filter out these formal first representations may seem excessive. In fact, it is normal and in the case of CDW27, given the complexity of the project and its ambitious targets, this time-scale and those that followed were remarkably tight.

The comments on model 4A, which began the series and

Making scale models in clay remains an important part of the design process, both for visual assessment and preliminary aerodynamic tests in the wind tunnel.

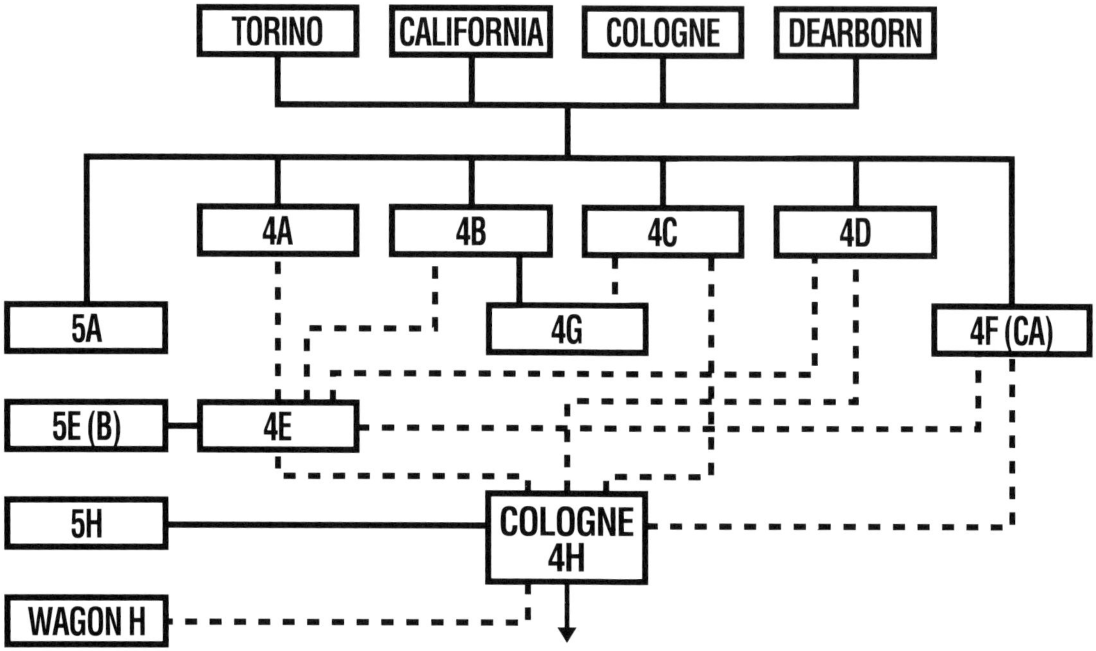

was completed in little more than three months, were short and to the point, as the objective here had been to minimize mechanical components and maximise interior space, emphasising the shift in proportions. It was considered well-proportioned with a good ratio of sheet-metal to glass. Boot proportions were good too, the perception of boot capacity being very important in the market segment that this car addresses. However, although it had very strong family appeal, this came at the expense of sportiness. Basically, after evaluating the car, it was felt that the shape was too balanced; the front and rear overhangs were too equal, it was lacking any novelty of proportion. And the tail-end, although meeting legal requirements, was not attractive and its blacked-out A pillar did not suggest interior spaciousness.

In November, the Merkenich studio presented model 4B in clay, followed in January 1988 by the glass-fibre version. This model was directly derived from the original Cologne target car, although the belt-line was simplified and the impression of boot capacity accentuated. Also, in comparison with the preceding four-light model, the glass-work was divided into six windows to give the impression of a more spacious interior. The major criticism centered on the fact that the car presented an over-pronounced wedge shape. In model 4C, constructed in clay in

The chart above shows the complex evolution of CDW27's design. The letters refer to specific models, which are referred to in the text and the pictorial spreads that follow.

the same month and developed in glass-fibre little more than two months later, positive elements came in the expressiveness of its front end, the dynamism of the side view and a very fresh overall design. Weak points were identified in its sporty nature, the complexity of surfaces, and the way in which the reduced dimensions of the back and greenhouse compromised the rear-most section of the roof.

In June 1988, the Dunton centre presented the fourth model of the first generation, 4D. It had been executed by their advanced design studio, again under the supervision of Lampe. Its greenhouse treatment was liked, as it gave a perception of good interior space. This was the first attempt at inset instead of the wrap-around 'limousine' doors; the Dunton car was felt, overall to be a good, sporty four-door.

The second generation of models was coded with the letters G, F, E and H, following the number 4 as before.

Proposal 4F came from the California Concept Center and had the precise function of evaluating further package changes. One side retained the regular wheelbase and on the other a

From "clays", the next stage in the design development is to make full-size models in glass-fibre.

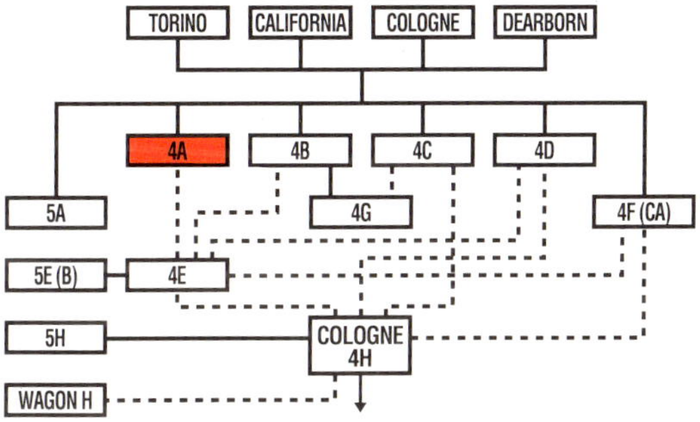

Full-scale models from the second stage, were of the four-door saloon version. The photographs here show 4A, from the Cologne studio. (As in the previous picture spreads of design concepts, the renderings on these pages are not necessarily directly related to the models shown.)

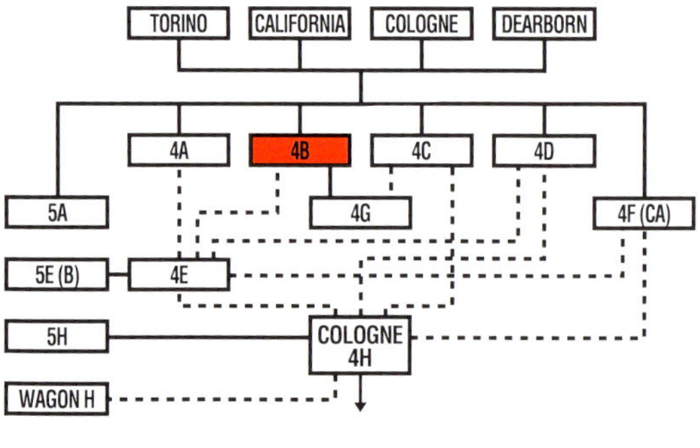
TORINO
CALIFORNIA
COLOGNE
DEARBORN
4A
4B
4C
4D
5A
4G
4F (CA)
5E (B)
4E
5H
COLOGNE 4H
WAGON H

Model 4B from Cologne was directly derived from the German studios original concept model.

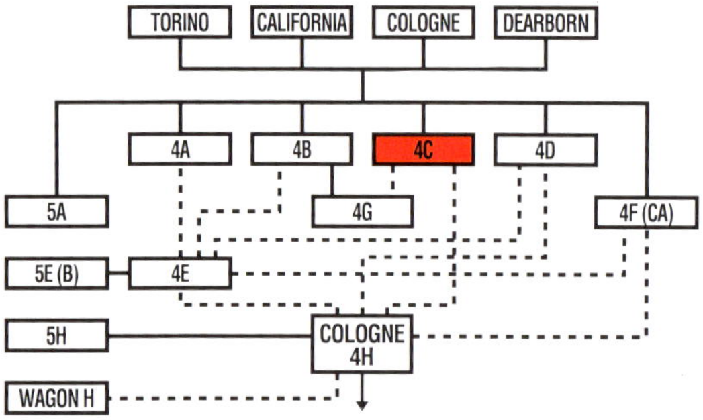
TORINO
CALIFORNIA
COLOGNE
DEARBORN
4A
4B
4C
4D
5A
4G
4F (CA)
5E (B)
4E
5H
COLOGNE 4H
WAGON H

4C, also from Cologne, was a more radical design, cut of the mainstream but which had some specific influences on the final version.

K - LN 000

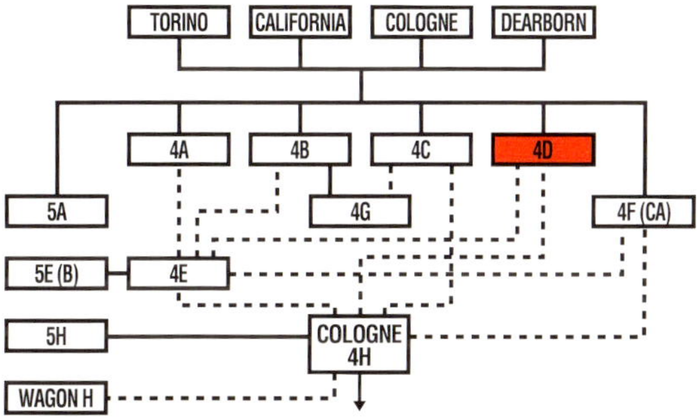

The advanced car studio at Ford's Engineering Centre at Dunton in south-east England created model 4D.

longer wheelbase and shorter rear overhang was experimented with, to improve access to the rear seats. 'But in the end,' recalls Lampe, 'we felt that the long wheelbase/short overhang section was too extreme and the other side was just a bit too normal.' Some of the ideas from the Californian model were later transferred to the Cologne model 4E, which also included elements from 4B and the Dunton model.

Proposal 4G from Cologne - which, chronologically, opened the second generation in January 1989 - had no apparent follow-up. In its four-light form some of the elements of model B and model C were integrated, but the treatment of the front end remained far from convincing.

At this point in the tale, Manfred Lampe explains that some of these exercises were not only concerned with styling: 'We also had to get a cost picture for the car, to assess if we had a business proposition, especially in terms of construction methods.' He then adds, by way of explanation, 'The Cologne, Dunton and California studies really contributed to the final model. Turin stopped with the initial concept model but the ideas it contained strongly influenced the first generation evolutions.'

At the same time, they took a first look at the five-door body, which was intended to be a more sporty looking car. If the four-door was more family-oriented, the five-door had to be rather more adventurous and have sportier cues. The five-door formed part of the CDW27 programme right from the start and the concept assumed that everything would be much the same, apart from the five-door's rear section.

Thus, in April 1988, there appeared a model code-numbered 5A, the scope of which was restricted to an assessment of whether the images set by the stylists' outlines could be achieved. 5A was used thereafter as a basic reference for further developments.

The final bodywork for the four-door started on the home straight with model 4E, defined in March 1989. This version is of fundamental importance. The designers, profiting from all that had gone before, made decisive final touches to the CDW27's architecture. In what is effectively a third generation, they left the overall dimensions unchanged but moved the rear wheels back by 50 mm. The front overhang was decreased by 25 mm, with the rear being lengthened by the same amount.

The form of 4E was reflected in the last model, 4H, and determined 5B (then coded 5E, being a development of the four-

Interior line-up - the many different proposals for CDW27's cabin were presented as seating "bucks", here shown with those who designed and built them.

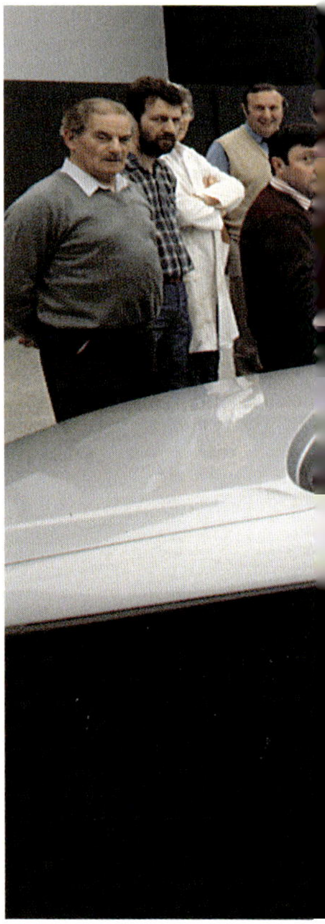

door), the model which, early in 1989, concluded formal research on the five-door bodywork and was finalized stylistically on model 5H in June of the same year.

So, it was model 4E that provided the references for development of the final form (4H) which was completed, albeit not definitively, in March 1989. The problem was the front. Doubts were raised about the integrated light-bar and the grille, inserted above a very dark lower area and an additional air intake. At the rear, the lighting strip was finalized, subsequent changes being restricted to colour toning.

The flanks of 4H expressed two different faces: one side featuring six-light glasswork and the other, four-light. It was decided in favour of the latter. But the most significant development was the disappearance of horizontal feature lines, giving way to a softer, more rounded form. The influence of recent Japanese styling, perhaps? 'Certainly not', insists Lampe, 'the decision to soften the sides was not the result of outside influence. This was our own assessment of where the style was

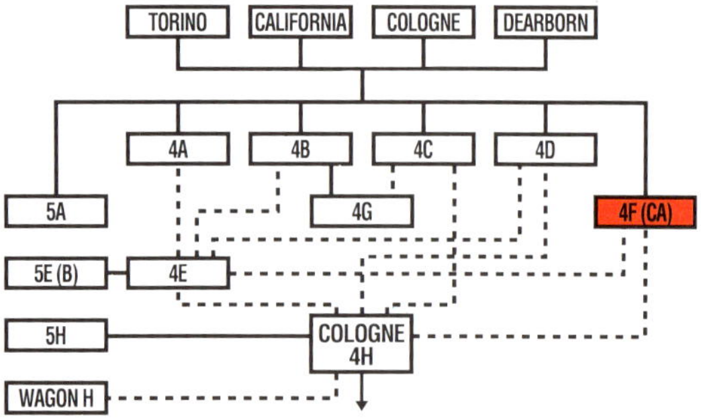
TORINO
CALIFORNIA
COLOGNE
DEARBORN
4A
4B
4C
4D
5A
4G
4F (CA)
5E (B)
4E
5H
COLOGNE 4H
WAGON H

STRONG 1·88

Proposal 4F, produced in California, explored package changes. Note the long wheelbase and very short overhangs front and rear.

K - LN 000

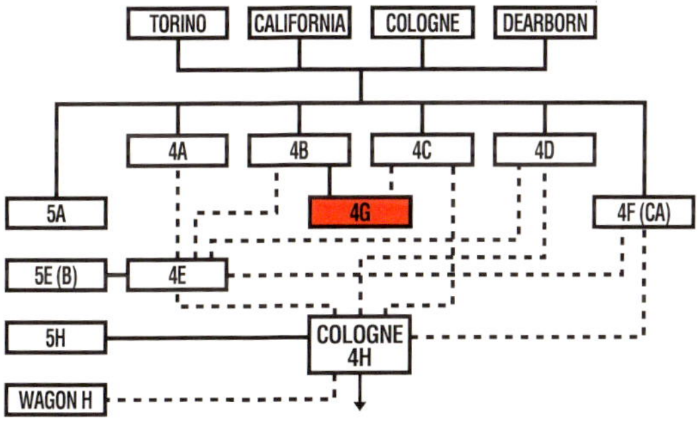

4G was a development by the Cologne studio from proposals 4B and 4C but did not progress further.

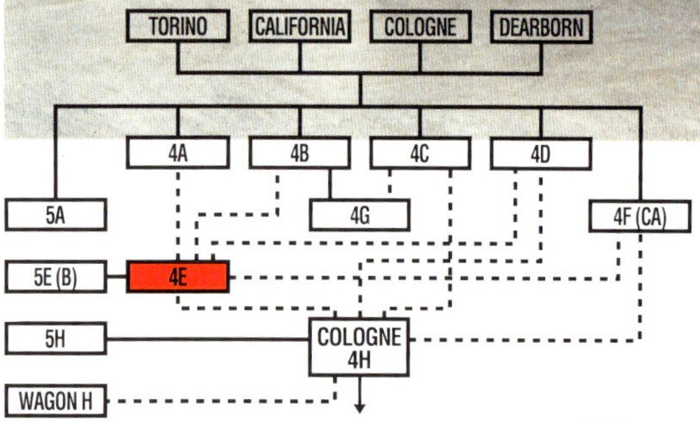

TORINO
CALIFORNIA
COLOGNE
DEARBORN
4A
4B
4C
4D
5A
4G
4F (CA)
5E (B)
4E
5H
COLOGNE 4H
WAGON H

Nearing the definitive shape - proposal 4E was the third generation design, completed in March 1989.

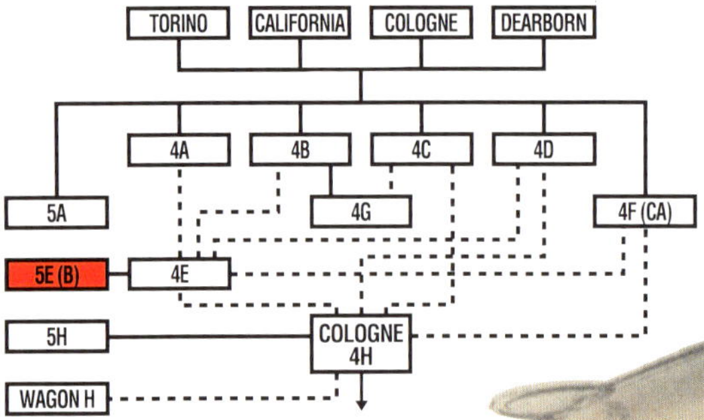
TORINO
CALIFORNIA
COLOGNE
DEARBORN
4A
4B
4C
4D
5A
4G
4F (CA)
5E (B)
4E
5H
COLOGNE 4H
WAGON H

This model, coded 5E or 5B, was directly related to the 4E four-door and set the style for the five-door Mondeo hatchback.

going to go. At that time we didn't know whether the Japanese would go in the same direction.'

For the design contingent, the style of CDW27 was settled in March 1989. The front and rear ends would be further refined after more aerodynamic tests to improve the drag factor and, more importantly, down-thrust.

Actually, work on the front end was to continue until May 1990. The vast documentation, drawings and clay models, of the forward section, developed under the supervision of Friedl Wülfing, testify to the effort required to arrive at a satisfactory solution. The final selection features an elliptical grille open in the lower part and dominated by a header bar dividing twin air intakes and a connecting bar carrying the Ford oval.

The style of the five-door was refined on model 5H but the differences were minimal compared with the previous version. The sheet-metal is common with the four-door with the exception of new roof, rear quarters, and, of course, the hatch. An important point is the unique tail lamps, which contribute to a highly differentiated look on the five-door. Also, the glass on the five-door covers the structure of the frame and further enhances the harmony of sheet-metal to glass, actually simplifying the shape.

In the opinion of Manfred Lampe and David Turner, who took Andy Jacobson's place in September 1989 as director of Ford Design in Europe, the Mondeo estate car (which will only be sold in Europe) is a 'terrific looking car', with its balanced proportions and pleasing rear end treatment. Dimensionally, the wagon is 50 mm longer than the hatch and saloon, while the wheelbase remains unchanged. Two design models were constructed. The first attempt, in June 1989, carried over the limousine door from the saloon with a steep inclination of the C pillar. In the second version, presented in October, the slope had been softened by several degrees, which left no choice but to design a new door.

The image projected by a car is determined by its overall design: inside as well as out. To complement a harmoniously-proportioned exterior form there must be a passenger compartment where 'living' conditions are the best possible, in terms of space, function, ergonomics, safety, finish, and versatility of use. In the CDW27 programme, the 'interior' design had a role equal to, and in certain respects more pronounced than that of the 'exterior' aspect.

The story of the evolution of the interior is, however,

simpler than that which led to the final outside shape and style. The development took place chiefly in the Dunton studios in England, working on three levels, coded E, N and M, representing the cars to be produced for Europe, North America (Ford Division) and Mercury. There was also, as with the exterior, a conceptual proposal executed by the Dunton studio and which was used in test clinics.

E1 had a modular pod-type control layout and a 'straight through' design with no driver orientation, with the intention of giving equal emphasis to both driver and passenger. It was then used to assess the cost-base, utilizing the appropriate materials. The theme of this mock-up was abandoned and the design direction changed for the next two proposals (E2 and E3). E2 was also developed for Mercury in the USA. These two were essentially along the same lines, with only one major difference: the fresh air vents of one were in the doors and in the other on

Even when the shape and design of CDW27 had been agreed, work continued on refinement of the front end styling.

Interior proposal N1, top, was for North America and, combined with European proposal E4, below, led to the definitive CDW27 cabin layout.

Design E1, left, was the starting point for the European fascia. Its "straight-through" design was rejected in favour of a "driver-orientated" module which was gradually refined and developed as shown in the illustrations below.

While instrument layout did not differ much between rival designs, ventilator position and steering wheel type were the subject of much discussion. The larger-hub steering wheel became inevitable when it was decided to standardize the air bag.

the instrument panel. Examined together, these two proposals were united in M3, which was a total package complete with airbags, had marked orientation towards the driver, control pods, door-mounted face-level air vents and clear stylistic continuity between instrument panel and console.

The E3 panel provided the basis for the eventual development of a version (E5) derived from Escort and Scorpio themes with a binnacle oriented towards the driver and the control area integrated within it. This prompted something of a déja-vu sensation however, in that it looked neither very new nor very advanced. It was instead the instrument panel designed for North America (N1) that originated the two mock-ups which led to the final definitive version. Another three models (M1, M2 and M4) evolved the theme and brought forth the panel destined for the Mercury-badged car.

The panel finally chosen offers a more curved form, has an airbag included in the package, a more prominent knee bolster, and one face-level vent incorporated into the command console while another is within the door. In the E4 mock-up, which anticipated the final form, the design of the facia and arched brow of the instrument panel enhance the design's fluidity and emphasise the sunken impression of the dashboard. Even the passenger-side air vent is inserted into the surface of the module.

In the definitive master model of the CDW27 interior, the close attention paid to ergonomics is clear. It is an ideally usable space and a friendly environment, with controls and instruments that are easy to use and understand. It is described as a 'touch and perceive' interior, capable of satisfying the most demanding of customers as well as meeting the requirements of the traditional buyer.

The interior style was actually finished before the exterior, in December 1988, and the styling of the dashboard was also considered valid for the US models, with the exception of certain details that are subject to different legislation.

Ray Everts, an American who now fills one of the Dunton design centre's key management positions and has long experience matured in the Ford studios at Dearborn and Cologne, underlines how the exterior and interior selected for the CDW27 merge harmoniously and invite the observer to perceive a unity of design: 'Everything for the interior was designed and developed in Dunton, even the fabrics with their original patterns and colours. These fabrics are going to be

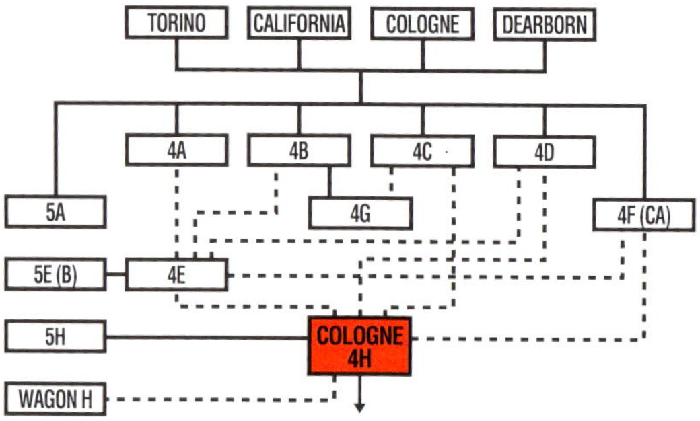

TORINO
CALIFORNIA
COLOGNE
DEARBORN
4A
4B
4C
4D
5A
4G
4F (CA)
5E (B)
4E
5H
COLOGNE 4H
WAGON H

Model 4H was the final design model, dating from March 1989, though further front-end changes were made over the following year.

STRONG 6.8.88

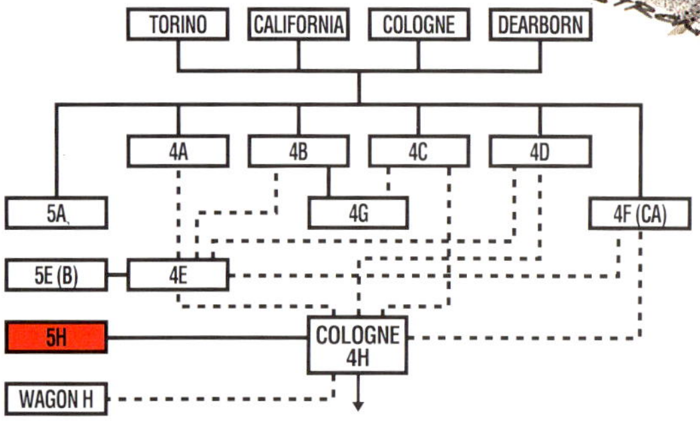
TORINO
CALIFORNIA
COLOGNE
DEARBORN
4A
4B
4C
4D
5A
4G
4F (CA)
5E (B)
4E
5H
COLOGNE 4H
WAGON H

5H is the definitive five-door, for Europe only, directly related to the final four-door model, 4H. Note the different rear lamp treatment compared with the saloon.

K- LN 000

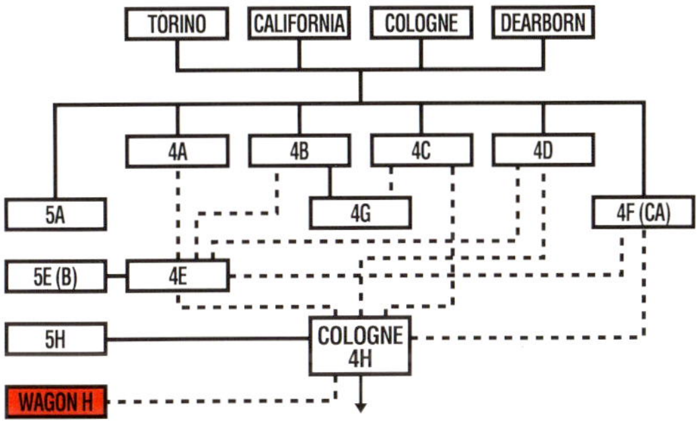

The Mondeo estate car, coded simply "Wagon H" by the design department, derives directly from the 4H saloon but needed unique rear doors to establish a harmonious shape to the rear quarters.

produced in the United States to our own specifications and will also be used on the CDWs built in the USA.'

The complete final design, inside and out, was finally approved at the end of 1990 by the design committee, which included Harold Poling, Ford's chief executive officer, and Jack Telnack, vice-president, design. But, in the case of CDW27, it is impossible to attribute paternity. David Turner, now director of design for Ford of Europe, and to whom Manfred Lampe and his managers reported in the programme's conclusive phase, sums it up: 'Credit must go to the entire Ford design team who, for four years, held nothing back in terms of creativity, passion and dedication in order to reach their true world-car target.' ●

The creative process combines with high-technology, as a clay model is digitized by a measurement probe, which will transmit the contours to a computer, where they will be stored and refined.

Clay scale model shaping up

FROUDE
0668
FROUDE
1929
rev/min
x 100
Tektronix

At the heart of Mondeo is Zeta, Ford of Europe's advanced new power unit, the sole four-cylinder petrol engine for CDW27. Prime mover of the US models will be a new V6 engine, also to be supplied to Europe from America. Together with a turbo-diesel, they meet all customer requirements - from economy to high performance.

AFFAIRS OF THE HEART

● **Arturo de Andrés**

At a quick glance, Mondeo's powertrain, its engines and transmissions, seem to be familiar stuff. It is a front-wheel drive car, with a transverse-mounted four-cylinder 'Zeta' engine and an MTX-75 five-speed transaxle.

But delve deeper into the development of CDW27, and look into the not-too-distant future, and there is much that is new. The Zeta engine appears in 2 litre form for the first time. Then, later on, an all-alloy 24-valve DOHC V6 of 2.5 litres will feative. An intercooled version of Ford's 1.8 litre turbo-diesel will become part of the line-up within a few weeks of launch.

There are also some important innovations in the transmission department: an electronically-controlled automatic four-speed gearbox, the CD4E, and a completely new four-wheel drive system.

Customer satisfaction is the key to what goes under Mondeo's bonnet. For Ford's powertrain engineers, this meant achieving class-leading levels of performance, excellent fuel economy, low exhaust emissions, quality, durability, reliability, and the kind of driveability which comes when performance, response and smoothness are harmonically blended. From the start a strong emphasis was put on reducing NVH (Noise Vibration Harshness).

These factors were overlayed by the global aspects of the CDW27 programme. While it was desirable that the line-up of engines and transmissions should be as simple as possible, there was also a need to offer a wide range of power and performance, and to meet present and anticipated exhaust emissions regulations in all markets.

Ford's new generation of engines was already planned when work started on CDW27. In the early 1980s it had become clear that the emphasis on efficient combustion (a result of the oil crises on the 1970s) would be tempered by the need to reduce exhaust emissions.

Ford's long-serving OHV four-cylinder 'Kent' engine had been gradually superceded by the Valencia-built HCS engine in the lower range of power, and by the single overhead-cam CVH from the Bridgend factory in the 1.3-1.6 litre segment. The 2 litre OHC engine gave way to the new Dagenham-built DOHC. There remained a gap between 1.6 to 2.0 litres - and that is where the new range, code-named Zeta, fits in.

To reduce range complexity - on both sides of the Atlantic - and aid the body design, it was decided at an early stage in the CDW27 programme that Zeta should be the only four-cylinder petrol engine to be used. At the same time it was recognized that top versions of the new car would eventually need more than the 136 hp (100 kW) offered by the Zeta 2 litre. The solution was found in an all-new 2.5 litre V6 designed in the United States.

This engine will become available 18 months after the car's European launch. Initially, Mondeo relies on Zeta engines, made at Bridgend in Wales and the 1.8 litre turbocharged four-cylinder diesel from Dagenham, England. In America, where CDW27 will come to the market some months later, the V6 will be available from the start and is expected to be the most popular powerplant. It is logical therefore, to source this engine exclusively from Cleveland, Ohio. The US range will include the Zeta only in 2 litre form and that will be built specifically for North American markets in the Ford factory at Chihuahua, Mexico.

By the middle of the decade Ford expects to be selling more engines with four valves per cylinder in Europe than all the Japanese manufacturers combined. Zeta production will rise to between 1.3 and 1.5 million units a year, at three sites - Bridgend, Cologne and Chihuahua.

Let's take a closer look at this engine, the heart of Mondeo.

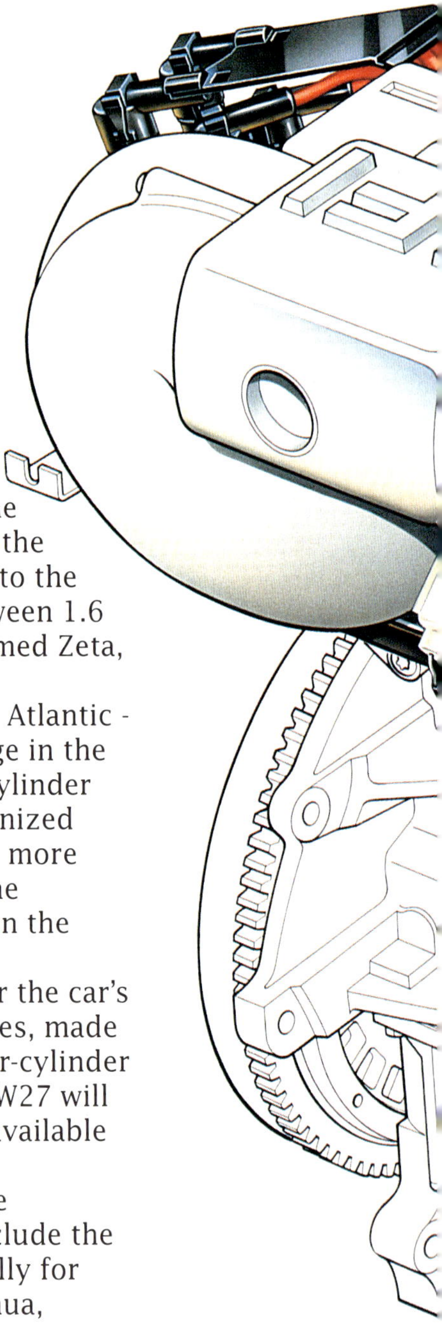

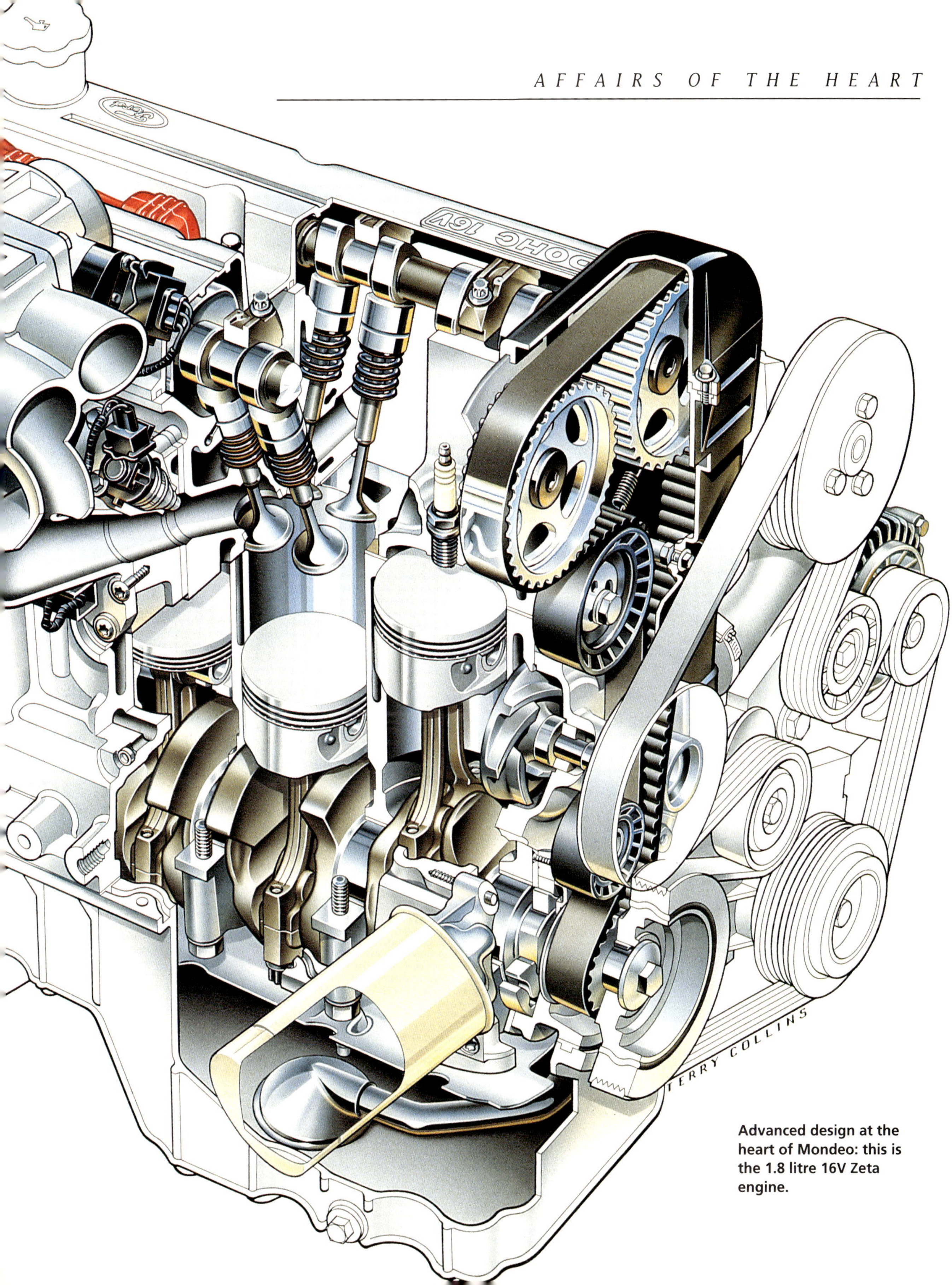

Advanced design at the heart of Mondeo: this is the 1.8 litre 16V Zeta engine.

It has been in service since spring 1992, in 1.8 litre form, powering some Escort/Orion and Fiesta models.

Zeta has an advanced design of cast-iron cylinder block fitted with a die-cast aluminium 16-valve, twin-cam head. The block shares a common 91.8 mm bore centre spacing with the CVH engine, in order to utilize existing manufacturing facilities. Grey iron sand casting was favoured for the block construction instead of aluminium because of its inherent advantages of stiffness and low noise radiation, and new design techniques ensured that there was almost no weight penalty.

It is offered in three displacements: 1.6, 1.8 and 2.0 litres. All three versions have the same 88.0 mm stroke. The shell-moulded cast-iron crankshaft has eight counterweights to reduce bending moments and improve dynamic balance. All main bearing and crankpin journals are fillet-rolled to reduce stresses at critical machining points. Forged steel connecting rods are teamed with lightweight, high-silicon aluminium pistons. In the more powerful versions these are cooled on their underside by oil jets.

In order to attain the maximum possible powertrain stiffness, there is a structural cast-aluminium oil pan, and the same material is used for the camshaft cover. To minimise NVH problems, advanced finite element modelling techniques were used to design the cylinder block and head, oil pan, manifolds, brackets and covers, and to optimize flywheel inertia.

In the CDW27 range there are as many as four engine mounts, keeping a constant radius around the engine torque roll axis; the stiffness of the body/chassis at the powertrain mounting points has been enhanced using CAE techniques.

These same techniques have been applied to the design of up to five resonators in the induction system, in order to absorb boom in the cabin and optimize performance.

In the exhaust system there is a new flexible joint between the manifold and the downpipe flange which lets the engine move on its mounts without allowing air to enter (which would alter the calibration of the emission control system), and without inducing high stresses in the main exhaust system. A laser technique was used to place the exhaust pipe hangers at points of zero vibration. The twin siamesed silencers - on all engines except the 1.6 litre - have been specially designed to minimize interior sound level, while keeping a soft but pleasing exterior exhaust note. Great effort was applied to reduce local noise generation from engine-mounted accessories and ancillary parts.

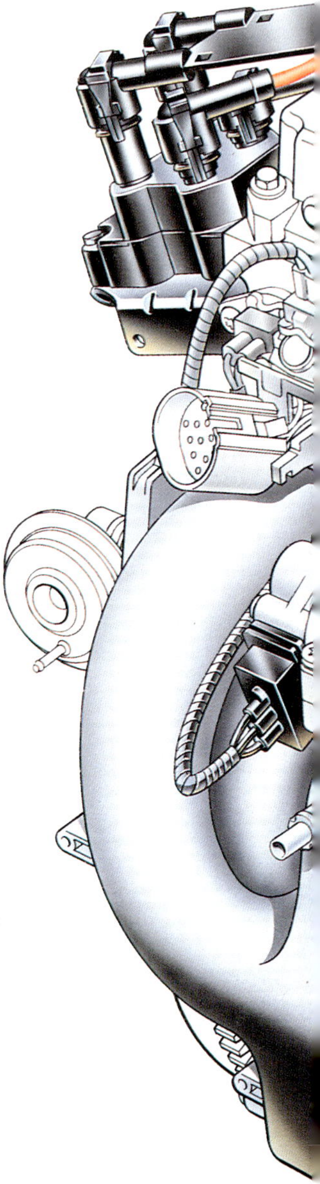

Powerful yet fuel efficient: the 2.0 litre version of the Zeta delivers 136 hp (100 kW).

Zeta's light-alloy cylinder head is cast from high-grade aluminium alloy, using metal external dies with conventional sand cores. Unlike Ford's previous applications of four-valves-per-cylinder technology, used to gain extra breathing capacity and hence more performance on cars like the Sierra Cosworth, Scorpio 24V, and Escort RS2000, the Zeta engine uses the four valves more as turbulence generator to provide better air/fuel mixing for faster, more stable combustion and lower exhaust emissions.

Four valves per cylinder allow the combustion chambers to have a smaller surface/volume ratio, lose less heat to the

cooling system, and the spark plug to be positioned at the centre of the construction chamber. After ignition, the flame front spreads uniformly through the fuel mixture, which burns more completely and efficiently, with less tendency to knock. This means it can tolerate either a higher compression ratio or a lower grade of fuel.

Of course, four valves provide the advantages of better engine breathing, reduced exhaust valve and valve seat temperatures, and less valve inertia, which in turn means reduced valve spring loadings, diminished cam wear, and higher speed potential.

But a four-valve combustion chamber can also do wonders for low speed torque; it is only a question of using conservative valve timing. This is easier to do than with a two-valve design, because the inherently better breathing allows the engine to be arranged with less valve 'overlap'.

In order to get smooth low-end running and low emissions, the Zeta engine was designed with a careful control of its combustion characteristics and high operating efficiency as top priorities. To achieve this it was necessary to study the gas dynamics phenomenon known as 'high velocity tumble', whereby highly efficient turbulence is generated at low engine speeds by arranging the port profiles so that the incoming gases naturally tumble into the chamber, without a significant loss of maximum flow capacity.

This improves the mixing of the fuel charge, and uses the squish generated towards the end of the compression stroke to accelerate the tumbling motion and cause it to decay into even more efficient micro-turbulent currents within the highly active charge motion. The result is clean burning of the fuel charge for stable idling and the best flame front propagation for complete combustion and maximum efficiency.

Zeta's valves are set at an included angle of 40 degrees and operated, via hydraulic bucket tappets, from twin overhead camshafts, machined from chilled-iron. Valve sizes are 32 mm for the inlet and 28 mm for the exhaust. An anti-syphon valve is fitted to prevent oil draining back from the oil galleries in the head when the engine is shut off, so that the tappets are not starved of lubricant. The camshafts each have five removable bearing caps, dowelled to the head for precise location; the bearings are located between the two cams for each cylinder, to spread the loads more evenly, and the fifth bearing resides behind the camshaft sprocket wheel. On some versions of the

A first in the Zeta engines used in Mondeo is the glass-fibre-reinforced intake manifold.

41 THRU 50 51 THRU 60
SMB-3B
E4LB-14A459-B
74004PC
74001MC
6128B-1861

Zeta engine, the cam lobes of each cylinder pair have slightly different profiles and are de-phased from each other by about 2 degrees, in order to separate the percussion peaks and reduce maximum loading on the camshaft.

The camshaft drive belt was the subject of a detailed study. It is made of highly-saturated nitrile (HSN) rubber, with glass-fibre reinforcement and a half-round tooth profile. The belt, 25 mm wide, is supported by a system of idlers which reduces the unsupported belt runs and minimizes the impact of standing shockwaves, a known source of premature belt wear. A special acid-treatment of the sprocket wheels provides an extra smooth surface finish. All this allows the service change interval for the timing belt to be extended from 60,000 km to 90,000 km.

The Zeta engine has been designed specifically to operate with a closed-loop three-way catalytic converter, and be fed by multi-point, sequential fuel injection. The four electronically-controlled fuel injectors are of the new side-fed type, and are pre-assembled into a forged aluminium fuel rail. These injectors, with smaller masses in their moving parts to allow a faster and more accurate response, employ fuel circulation inside their main bodies, making them much less sensitive to high temperatures; this helps to provide immediate starting under the widest range of operating conditions.

A first in the Zeta engines used in Mondeo is the four-branch tuned-length intake manifold built in glass-fibre-reinforced nylon resin. This gives a 3.5 kg weight saving compared with aluminium, and the fine surface finish improves the air flow to the extent of a 2 per cent gain in torque. Starting the engine in very hot conditions is considerably easier due to the thermal characteristics of this manifold, which helps to reduce the temperature of the injectors.

Ignition is of the electronic, programmed and static kind, distributorless and with two back-to-back coil assemblies. Both the ignition and the sequential fuel injection are controlled through a new version of Ford's widely-used 16-bit EEC-IV engine management computer, whose module has been uprated from 32 to 56 kB capacity, and its controller clock speed increased from 12 to 18 MHz. The Zeta is the first Ford power unit to incorporate this version of EEC-IV, which is capable of handling information at the rate of 1.25 million instructions per second.

In addition to the established Ford STAR tester interface,

Electronic heart - the module of Ford's widely-used 16-bit EEC-IV engine management computer has more capacity and increased clock speed.

this latest version of EEC-IV is provided with a new data communication link, which allows the module to be interrogated directly, to check the function and integrity of its components.

Idle speed is controlled independently of the main throttle plate by a separate small-bore, fast-acting valve, in order to minimize speed variations produced by changes of load due to torque-consuming devices such as power steering and air conditioning.

For crankcase ventilation, special attention has been paid to prevent the formation of oil sludge: an automatic valve assembly is fitted in order to vary the ventilation air flow to the crankcase in proportion to the blow-by pressure, which in turn depends on load and throttle opening. Thus, the crankcase is always under negative pressure, so oil fumes are prevented from escaping and fed to the air intake.

During the warm-up after a cold start, an exhaust air injection system uses one-way valves to inject secondary air

Engine warm-up from a cold start is directed by EEC-IV module and uses valve system to inject secondary air into the exhaust and promote catalyst efficiency.

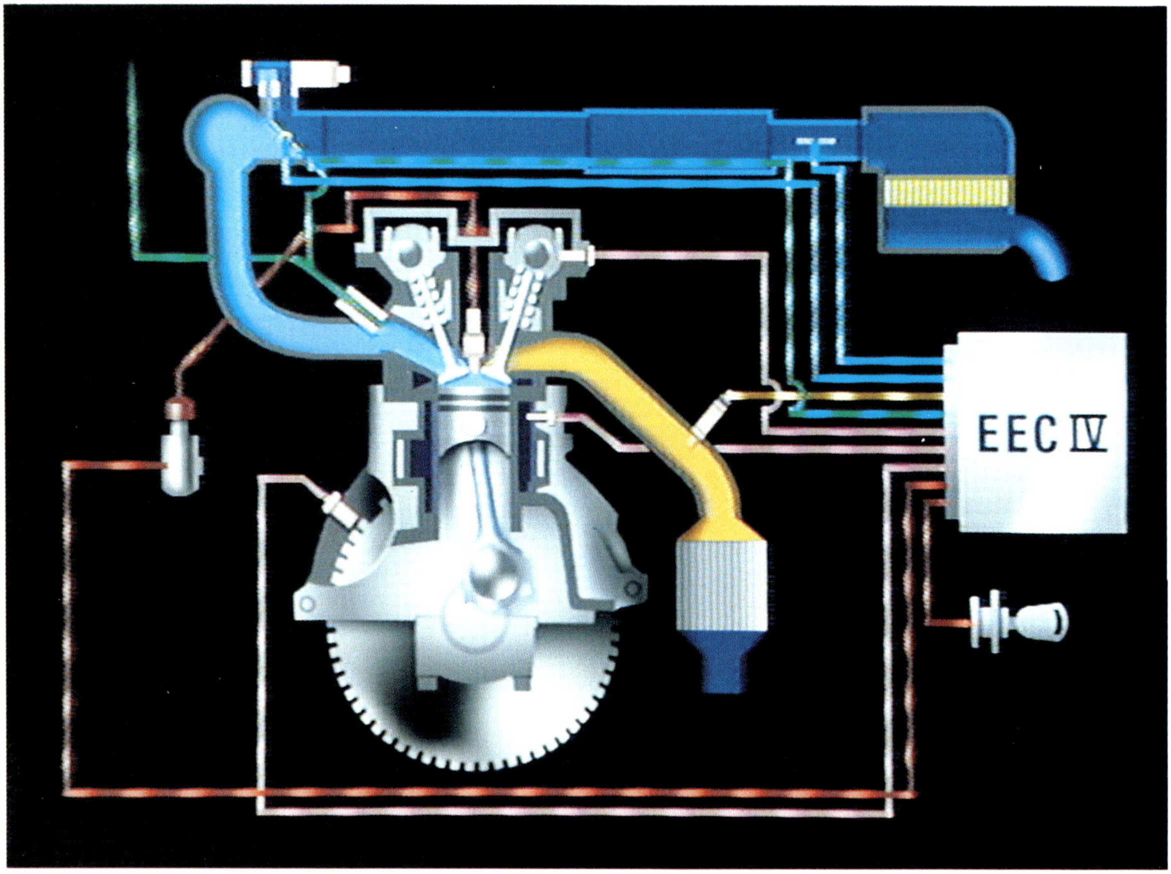

into the exhaust manifold, using the flow resonances to generate suction pulses; the excess oxygen contributes to oxidizing the hydrocarbons and carbon monoxide gases, and promotes a rapid rise in the catalyst temperature. As soon as the engine is warm, the pulse air is switched off by the EEC-IV module, to enable the closed-loop controls to take over.

The catalytic converter is placed at the optimum position for rapid warm-up from a cold start, but not so close that it presents a risk of de-activating the catalytic coating from excessive heat build-up at high speed. The catalyst chamber has a large cross-sectional area to minimize back-pressure.

The engineers are very proud of Zeta's quick and reliable starting. The target, under normal conditions, was just 0.5 second; that is, turning the key and releasing it without waiting at all. This was not an easy goal, because the engine needs to begin turning before all the electronics for injection and ignition can start to work, a sensor on the flywheel being the trigger for the exact moment of ignition spark and fuel injection. With the development of the EEC-IV module, it was achieved, and even at minus 15 degrees C, the starting time is no more than one second.

Zeta engine started its life as a 1.8 litre engine: 1,796 cc with a bore and stroke of 80.6 mm and 88.0 mm, compression ratio 10.0:1. For the Fiesta and Escort applications there are two versions: 105 and 130 hp (77 and 96 kW). These have different camshafts, injection and ignition settings, the 105 hp engine emphasizing torque and the 130 hp, higher speed.

For Mondeo, the settings are different again, and fall more or less in the middle of the two existing versions, but with a bias towards mid-range torque. This third version of the 1.8 Zeta develops 115 hp (85 kW) at 5,750 rpm and 158 Nm at 3,750 rpm.

To put this in perspective, the Zeta 1.8's figures would be deemed perfectly adequate for an 8-valve 2.0 litre engine. Of particular significance is the fact that 90 per cent of maximum torque is available between 1,800 and 5,700 rpm.

On the other hand, the engine's rev limit is very high; continuous running is allowed up to 6,450 rpm, with intermittent bursts to 6,675 rpm. In fact, all Zetas are remarkably free-revving, capable of running smoothly to 7,000 rpm, whereupon a 'soft feel' limiter begins to operate by progressively cutting the fuel to each cylinder.

The economy-minded owner can choose a Mondeo with the

new 1.6 litre version of the Zeta: 1,597 cc with bore diameter reduced to 76.0 mm, compression ratio 10.3:1. The idea here is to obtain as much low-end and mid-range torque as possible - as the figure of 139 Nm at 3,500 rpm indicates - but the mild camshafts don't encourage heavy breathing at high revs so the maximum power is a conservative 90 hp (66 kW) at 5,250 rpm.

At the other end of the Zeta scale is the 2 litre: 1,988cc, bore 84.8 mm, compression ratio 10.0:1. Its performance is really impressive, with 136 hp (100 kW) at 6,000 rpm and 180 Nm at 4,000 rpm. For this one, all the settings favour the middle and high range, and although 90 per cent of the maximum

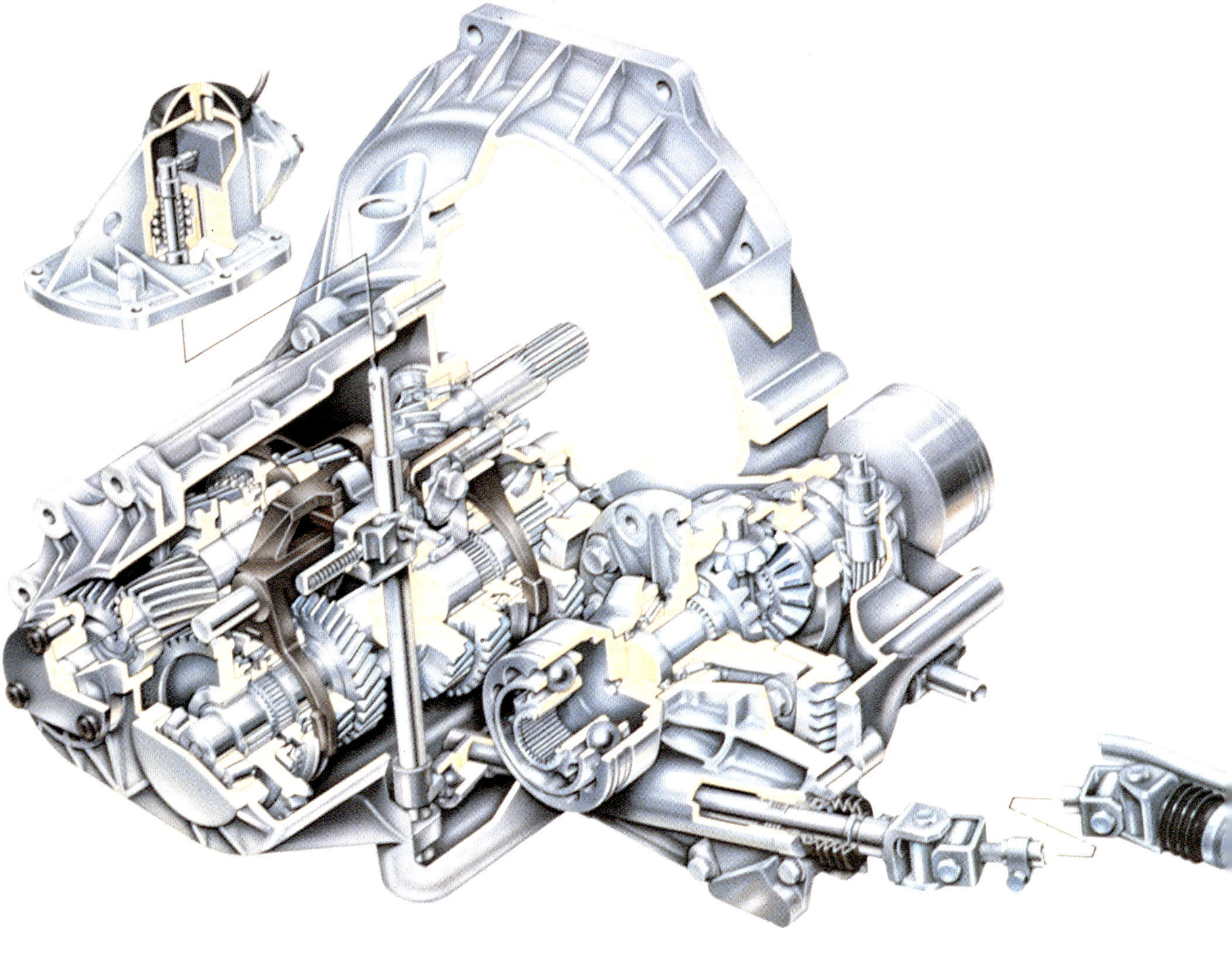

torque is available over a wide span of revs, it is biased more towards the top end, between 2,400 and 6,050 rpm. The rev limits, for continuous and intermittent use, are the same as for the 1.8 litre.

It is interesting to compare the 2 litre with the 1.8 litre, which is the central and reference version of the Zeta engine. The regimes for maximum power and torque go up 250 rpm in each case, but the absolute performance of the 2 litre is better; in direct proportion, it would have no more than 127 hp (93 kW) and 174.5 Nm.

A turbo-diesel Mondeo will be available within a couple of months of launch. The four-cylinder diesel engine is currently in use in other Ford products, but not in the form used here, as for the first time the turbocharger is accompanied by a charge air intercooler.

With a bore and stroke dimensions of 82.5 mm and 82.0 mm, the 1,753 cc diesel is a 'square' design. Both cylinder block and head are cast-iron, a classical and safe technique for diesel engines. Camshaft drive is a mixed system, with gears to the injection pump and a cogged belt to the overhead camshaft. The geometrical compression ratio is 21.5:1.

Turbocharged diesels particularly appreciate the fitting of an intercooler; fitting one to the Mondeo's diesel boosts power from 75 hp (55 kW) of the previous turbo-diesel used in the Sierra, to 90 hp (66 kW), and maximum torque goes as high as 183 Nm at 2,250 rpm, with 90 per cent of this figure between 1,570 and 3,670 rpm.

This engine has the same torque as with the 2.0 litre Zeta, and with much better low-end punch - but it is also required to pull much higher gearing. The development team have given special attention to some quirks of turbo-diesels, seeking not only to eliminate turbo lag in the response to the throttle, but also the sudden arrival of 50 Nm of torque in the narrow band of revs between 1,300 and 1,800 rpm.

The all-aluminium 24-valve DOHC 2.5 litre V6, that will come a little later in Mondeo's career, is an absolutely new engine. It will be supplied from Cleveland

MTX-75 five-speed manual transmission is compact but has high torque capacity.

for all the world, and in the same specification. Its block, heads and sump are made in aluminium. Each cylinder head has two camshafts, operating four valves per cylinder. Bearing in mind that it will meet the tough 1994 US exhaust emissions regulations from the outset, the performance of this engine is remarkable. It produces 171 hp (126 kW) at 6,000 rpm, with a maximum torque of 222 Nm at 4,500 rpm.

Just as most US versions of CDW27 will be V6-powered, so will the majority there have automatic transmission. That takes us to the last part of the new Ford's powertrain story.

Mondeo has two kinds of gearboxes, five-speed manual and four-speed automatic. It also offers front-wheel drive and four-wheel drive. But not all combinations of engine and transmission will be available. Automatics will go only with front-wheel drive, and exclusively with V6, 2.0 and 1.8 petrol engines. The Mondeo 1.6 and 1.8 turbo-diesel will be only front-wheel drive with manual transmission. The V6 can be either manual or automatic, but not four-wheel drive. That leaves the 2.0 litre and the 1.8 Zeta as the engines for the optional four-wheel drive, with manual gearbox.

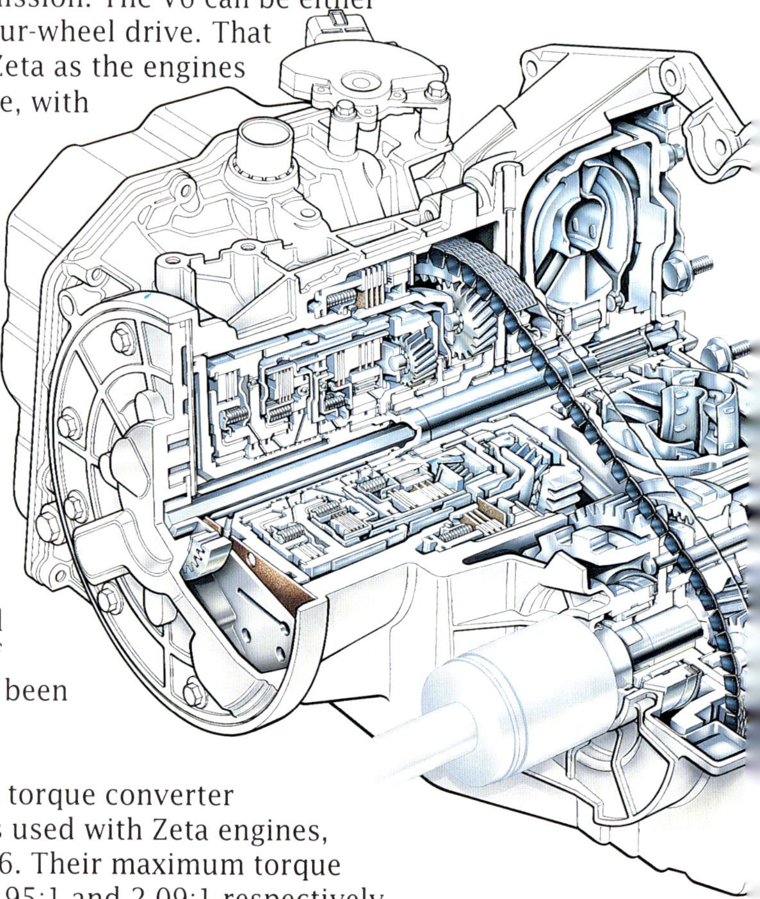

The automatic transmission, code-named CD4E, was developed in USA, and makes its debut in Mondeo. Its control is fully electronic, the torque converter is partially by-passed in third gear, and fourth is a totally mechanical overdrive. It offers, at the touch of a button, the choice of 'economy' and 'sport' driving modes and, for the first time, all the controls for the operation of an automatic transmission have been integrated with the EEC- IV microprocessor.

Diameters of the hydraulic torque converter is 9.25 in (235 mm) for the units used with Zeta engines, and 10.25 in (260 mm) for the V6. Their maximum torque increases at stalling speed are 1.95:1 and 2.09:1 respectively.

Gear ratios - in both cases, 2.890; 1.571; 1.000; 0.698 - are fairly wide, in order to assure a brisk starting acceleration and economical cruising.

Mondeo's four-wheel drive system is novel. Although not as complex a system as the 4X4 layout of the Sierra and Escort Cosworth, with their two Ferguson viscous couplings and 34/66 front/rear torque split its 58/42 torque split is combined with, and a traction control system to cut the power when the ABS sensors detect that a wheel is slipping.

In Europe, most Mondeos will be sold with manual gearboxes and two-wheel drive. Until now the MTX-75 gearbox has been used transversely only in the Escort RS2000 and the Escort and Orion with the 1.8 litre 130 hp (96 kW) Zeta engine.

Made in Halewood and Cologne, MTX-75 stands out because of its high torque capacity of 270 Nm while still being lightweight, thanks to its two-piece die-cast aluminium casing. With closed ends, minimal external seals and top-mounted shift mechanism, it is designed to minimize the risk of oil leakage.

To keep the shifting effort low, the two-rail linkage is supported on multiple ball-race bearings, and pre-assembled before installation. And to achieve a quieter running and less power losses, all gears run on needle roller bearings. Low-viscosity automatic transmission fluid (ATF) is used and the gearbox is filled for life.

The MTX-75 is built with high standards of precision. Its denomination derives from the 75 mm between its two shafts, distance that is kept to the very close tolerance of plus/minus 0.015 mm. The production system (which is the same for the longitudinal MT-75 used in Sierras and Granadas) incorporates a laser-welded two-part gear and synchronizer, whose cones are temporarily joined for hardening, before being separated for fine honing of the helical profiles and then laser-welded permanently together.

There are three different sets of gears for the CDW27, the most popular being the one used in the 1.6, 1.8, and V6. These ratios are: 3.417; 2.136; 1.448; 1.028; 0.767. The first-to-fifth span is 4.455:1 - a typical touring car gear set.

The turbo-diesel has much wider ratios, that engine's elasticity, and relatively low horsepower allowing a greater span of 5.430:1, from a low first gear to a high fifth, for economical cruising. Its ratios are: 3.660; 2.047; 1.258; 0.864; 0.674.

The arrangements for the Mondeo 2.0 are somewhat different. Here, a semi-close ratio gearbox is preferred, as befits

The four-speed automatic transmission, code-named CD4E, newly developed in the USA, makes its debut in Mondeo.

a more powerful and sporting engine. So, the gear ratios are much closer: 3.231; 2.136; 1.483; 1.114; 0.854. The first-to-fifth span in this case is only 3.783:1 - more or less the same jump as first-to-fourth in the 'touring' gear set.

Even a powerful engine in a light and streamlined car can see its performance and/or fuel economy ruined by the wrong final gearing. For CDW27, final-drive ratios have been carefully chosen for each model. The Mondeo 1.6 has a 4.060 final drive ratio which, with 185/65-14 tyres, gives a final gearing of 16.3 (26.2 km/h) and 21.8 mph (35.1 km/h) per 1,000 rpm in 4th and 5th gears. Its maximum speed is 112 mph (180 km/h), 0-100 km/h (62 mph) acceleration, 13.5 seconds.

The same gearing and tyres are used with the 115 hp (85 kW) 1.8 engine, but the increased power gets this version from 0 to 100 km/h (62 mph) in 11.0 seconds, and up to a maximum speed of 121 mph (195 km/h). There is also the option of 195/60-14 tyres, with a slightly smaller rolling diameter, and with those the final gearing goes down to 16.1 mph (25.9 km/h) and 21.5 mph (34.6 km/h) per 1,000 rpm in fourth and fifth gears.

Those same tyres, also with the 4.060 final drive ratio, can be had on the 2 litre Mondeo 'sport' but, because of the different gearbox, the final gearing is even shorter, at 15.0 mph (24.1 km/h) and 19.6 mph (31.6 km/h) per 1,000 rpm in fourth and fifth.

It is clear that, 'sport' version or not, the 2 litre Mondeo is intended to have a sporting character. Look at the performance figures: maximum speed is 127 mph (204 km/h) and 0 to 100 km/h (62 mph) is attained in 9.6 seconds. Naturally, fuel

FORD'S MODULAR ENGINES

Mondeo's American V6 is part of a new family of Ford engines in vee configuration, with the number of cylinders ranging from six to eight, and even 10. It is a modular design, in which not only the number of cylinders, but also the angle between them - 60 or 90 degrees - and the construction material, can be changed.

The 60-degree engines will be built in the factory at Cleveland, Ohio, while the 90-degree versions will be made at the Romeo, Michigan plant. Such is the flexibility of this modular concept, that the cylinder heads can be of the SOHC or DOHC type, with two or four valves per cylinder, and the base material either cast-iron or light alloy.

The range of cylinder capacities inside this family of engines will be enormous, from a 2 litre V6 to a 5.4 litre V8 - with a V10 kept in the reserve. Mondeo will have a 2.5 litre 60-degree V6.

Up to now, the only versions offered for sale are the 4.6 litre 90-degree V8 in the Lincoln Town Car, Mercury Grand Marquis and Ford Crown Victoria and the 32-valve engine in the new Lincoln Mk VIII.

consumption is higher than the smaller and longer-geared engines but that is the trade-off for a 136 hp (100 kW) engine and a top gear of under 20 mph per 1,000 rpm.

For the V6 model with a manual gearbox, the final drive is 3.820 but with 205/55-15 tyres and a return to the 1.6/1.8 gearbox ratios. That gives 17.6/23.5 mph (28.3/37.8 km/h) final gearing, and an impressive top speed of 137 mph (220 km/h). Its acceleration is good too - 0-100 km/h (62 mph) in 8.6 seconds.

The turbo-diesel has a different role in life. Though it keeps the short 4.060 final drive - and 185/65-14 tyres - its longish fifth gear pulls the final gearing to 19.3 mph (31.1 km/h) in fourth and 24.8 mph (39.9 km/h) in fifth. Performance, of course, returns to relatively modest levels, but keeps on par with the 1.6 Zeta; acceleration is, at 13.4 seconds to 100 km/h (62 mph) a little slower, due to the greater inertia of the diesel's rotating parts, but the top speed is marginally better, at 114 mph (183 km/h). As expected, its fuel economy is exceptionally good for a car of this weight and size - better than 66 mpg (4.3 litres/100 km) is achieved at a constant 56 mph (90 km/h).

With automatic transmission, both the 1.8 and 2.0 Zeta engines share a 3.916 final drive ratio, with a final gearing of 24.8 mph (39.9 km/h) in top. That gives the 1.8 0-100 km/h (62 mph) acceleration in 13.5 seconds and a maximum speed of 111 mph (179 km/h) and the 2.0 corresponding figures of 11.6 seconds and 121 mph (195 km/h). Fuel economy suffers less than might be imagined, thanks to the electronic management of the transmission and the mechanical by-passing of the torque converter in the higher gears. The mpg figures for the 2 litre automatic are actually slightly better than the manual version with the same engine and a manual gearbox.

Finally, the automatic V6, which will be the most popular CDW27 in the USA, features a 3.496 final drive ratio, has 205/55-15 tyres, and really long gearing: third, 19.7 mph (31.7 km/h) per 1,000 rpm and fourth, 28.3 mph (45.6 km/h). The 0-100 km/h (62 mph) figure goes up to 10.5 seconds, and top speed down to 127 mph (205 km/h) - still fast enough to get speeding tickets in almost any country in the world! ●

Technology has redefined every aspect of car design and development. It took a year's calculations by Ford's supercomputer to achieve the optimum combination of safety, strength, and weight for CDW27. Achievement of the highest standards of safety - both passive and active - underlined the whole programme.

OBJECTIVE: SAFE AND SURE

● **Jürgen Lewandowski**

It's not easy to pursue a technical career: people never stop telling you how they would have designed and built a better, more economical and environmentally acceptable product at an unbeatably low price. Customers, journalists and professional colleagues could all have solved your problems in a flash, given unlimited funds and more patient company executives.

Automobile engineers face a particularly difficult challenge. It takes several years before the decisions they take turn up as sparkling new sheet metal in the showrooms. Many years later, the last specimens of their creative genius are still to be seen on our streets. In the meantime, they are busy putting the finishing touches to their latest models, only to find their most brilliant technical concepts anticipated by a competitor who reaches the market six months before them.

A completely new design is like taking a dive off the top board. Ford's engineers faced this situation five years ago when they drew up the initial specification for CDW27. Their task was made more difficult by the requirement for this to be a global car, built initially on both sides of the Atlantic for markets all over the world.

The floorpan and front-end structure had to be identical for every version, though it was accepted that the body itself could be varied to suit local market requirements. The European car had to be right- and left-hand drive. And since the US version might be exported to Central and South America, where it could encounter less civilized driving conditions, the car's suspension would have to be exceptionally rugged.

Ford's marketing experts say that it is no longer appropriate to talk about 'typical' European and American customers. Since a car has to perform equally well in the desperate cold of a Finnish winter and the searing heat of Arizona, they talk now of the 'North Atlantic customer'. CDW27 was to be the first new car aimed specifically at satisfying the expectations of this wide customer group.

In due course, contributions to the new car's specification were received from the various company divisions: Design, Vehicle

Engineering, Body and Chassis, Electrical, Engine, Transmission, Parts and Service, Manufacturing, Purchase and Marketing. Their market analyses, long-term forecasts and suggested working methods all went into the draft specification.

Mondeo cutaway shows compact mechanical layout, suspension sub-frames, and safety beams in the doors.

The heads of these divisions drew up a list of the following 13 parameters:

'RIDE'	• the best possible suspension for both rough surfaces and high-speed driving.
'HANDLING'	• would the new car be easily controllable by drivers with no Grand Prix ambitions?
'STEERING'	• combining low effort and precision is an art.
'POWERTRAIN'	• engine and transmission had to match one another perfectly.
'BODY/CHASSIS NVH'	• top design and build quality was essential for the basic structure.
'DRIVEABILITY'	• the new car should be easy to drive and problem-free in use.
'SHIFTABILITY'	• a precise, light-action gear shift, another far-from-easy task.
'BRAKE FEEL'	• the brakes should slow the car firmly without grabbing.
'OPERATIONAL COMFORT'	• all controls to be well placed and within easy reach.
'SEAT PERFORMANCE'	• are the occupants still comfortable, with no back pains, after a 500 mile journey?
'CLIMATE CONTROL'	• will driver and passengers feel happy in the car despite severe outside temperatures?
'ILLUMINATION'	• ideal lighting inside the car, and on the road ahead.
'PERFECTION FEELING'	• possibly the most difficult category of all: the car should not become unreliable or lose its feeling of build strength over the years.

All this left one major question open: by what standards should the new car be assessed? Whose task was it to lay down a basis for comparison?

Technical people think in real terms, and that meant comparing their new concept with real cars - those already on the market and praised for specific features of their design. Of course, the most recently-launched models are bound to have a certain technical advantage. When the Ford Sierra first appeared, other manufacturers analyzed it, right down to its component parts. Now, as they prepared to re-define the state of the art with CDW27, Ford's engineers had to take a hard look at what their rivals had produced in the meantime.

After this intensive scrutiny of the competition, the Ford team set standards for CDW27 in each of the 13 categories. Their mission was to make it the best in its class - and if we are to take their word for it, they were successful.

It is said that, to be among the leaders in chassis design today, two things are needed: a giant 'supercomputer' and a team of engineers with the intelligence to program it.

Chassis and and body design are the key elements in any new car concept. It is not easy for an outsider to understand the complexities of chassis design, nor how the specialists involved go about their job.

Consider this: When a car wheel runs over an uneven surface at high speed, it moves up and down so violently that the suspension is unable to absorb all the resulting oscillations. Some of them are transmitted to the body. We are accustomed to some body movement occurring even on cars described as having a 'flat' ride or 'cornering on rails'. Furthermore, all four wheels encounter different irregularities of the road surface, so with independent suspension all round, there are four points at which the unfortunate bodyshell is pushed, pulled and twisted. The resulting distortion makes it very difficult to maintain the chosen suspension and wheel alignment settings.

The more rigid the bodyshell, however, the less twisting movement occurs and the easier it becomes to maintain the correct suspension settings. A rigid bodyshell has other advantages, not least in terms of passive safety for the occupants. If the passenger section of the body is strong and rigid, the crumple zones at the front and rear can absorb collision impacts and the occupants stand a much better chance of emerging from an accident safe and uninjured.

These days the way to arrive at the optimum body

Expert advice: former world champion F1 driver Jackie Stewart with CDW27 prototype.

structure is by FEM - Finite Element Method - and to make those calculations on a complete car body requires a truly gigantic machine, like the Cray 4MP supercomputer at Ford's Dearborn headquarters.

FEM involves breaking every single square millimetre of the body down into a multitude of small geometrical units - tiny squares that are analysed for their deformation when a load is simulated, and for the way they affect adjacent areas of the body. For certain structural analyses, the computer has to examine as many as 70,000 finite elements at the same time.

The engineers were pleased to find that seven of the 11 points on the CDW27 body that they knew would be subjected to the most greatest stresses did in fact fracture when subjected to the most severe endurance testing on real prototypes. This confirmed the accuracy of the FEM work. Strengthening the body at critical points becomes a much easier process after this kind of thorough analysis, and the engineers can strike a balance between body rigidity and extra weight, without the expensive and time-consuming tests that used to be required.

If all the time spent on calculations for the CDW27 body by the mighty Cray computer is added up, it comes to about a year!

Just compiling the necessary programs took between 3,000 and 4,000 hours.

When the basic design of the CDW27 body had been finalized, engineers in Cologne embarked on construction of the first prototypes. They also built a 'Royalite' model. Royalite? This is Ford's term for the transparent plastic models it produces for each new generation of new car. Their purpose is to show to all concerned just how the new body has been designed and how it will be put together on the assembly line.

The computer generates a three-dimensional outline for every single body element. This is converted into a numerically controlled milling-machine program, so that the pattern-making shop can produce a wooden or plastic pattern for the panel or

Hi-tech testing - fully-instrumented CDW27 prototype at the Lommel proving ground.

NO-COMPROMISE STEERING

Mondeo has power steering as standard throughout the range. Only a couple of years ago, such a move would be unthinkable in the CD class. The reason for it was not only to ensure that customers could park and manoeuvre without undue effort but also to avoid compromise in the car's handling.

CDW27 programme director David Price recalls that originally the steering castor angle was 1.7 degrees, which was thought to be the optimum for a car that would have assisted or non-assisted steering. 'Then we had some very specific drive appraisals aimed at castor angle changes and found that with 1.7 degrees we were getting fish-tailing that we didn't like and that the handling was never going to come up to expectations. It needed 2 degrees - which meant standardizing power steering. That was a big decision at the time but I am really glad we did it.'

Like the automatic transmission, and V6 engine for 1994, the steering system for CDW27 is sourced in the USA, from Indianapolis. It has three turns from lock-to-lock and the power assistance is linear. An unusual feature is a high frequency valve that checks rapid transfer of fluid within the system and is designed to prevent the steering 'fight' that afflicts some front-wheel drive cars when accelerating hard.

stamping concerned. Even at this stage, the surfaces are accurate to within one-fifth of a millimetre. Small holes are drilled in these patterns and plastic film applied to their surfaces. They are heated and air extracted through the holes, so that the plastic film takes on the precise shape of the pattern beneath. The resulting plastic mouldings can then be tacked or clipped together to yield a most convincing - though rather fragile - replica of the finished body.

All this sounds more complex than it actually is, but the biggest advantage is that everyone can obtain a clear, three-dimensional picture of a complex modern car body. The production experts can check that their robots can reach all the welding points; the after-sales and service people can satisfy themselves that various mechanical assemblies are easily accessible.

With this degree of technology and pre-production analysis, it was predictable that the real CDW27 would do well in its crash tests. At speeds up to 15 km/hr (9.3 mph), the main side members deform by a maximum of 30 millimetres, but do not make contact with the engine. Even at still higher collision speeds, the body's crumple zones absorb impact energy admirably well, and the occupants are exposed to loads lower than any achieved by existing models. This is a decisive passive safety factor - and one that easily repays the months of work on one of the world's most powerful computers.

Suspension design also has an important bearing on vehicle safety. In this case, the motto for CDW27 was 'evolution not revolution', and it was decided to retain many of the features that Ford had developed to a high level of perfection over the years. Unusually for a car in this category, both front and rear suspension are mounted on separate sub-frames. This layout has three significant advantages: vibration from the wheel locating elements can be largely isolated from the main bodyshell; the running gear and the bodyshell can be made as strong as possible; and both initial assembly and subsequent repair are simplified.

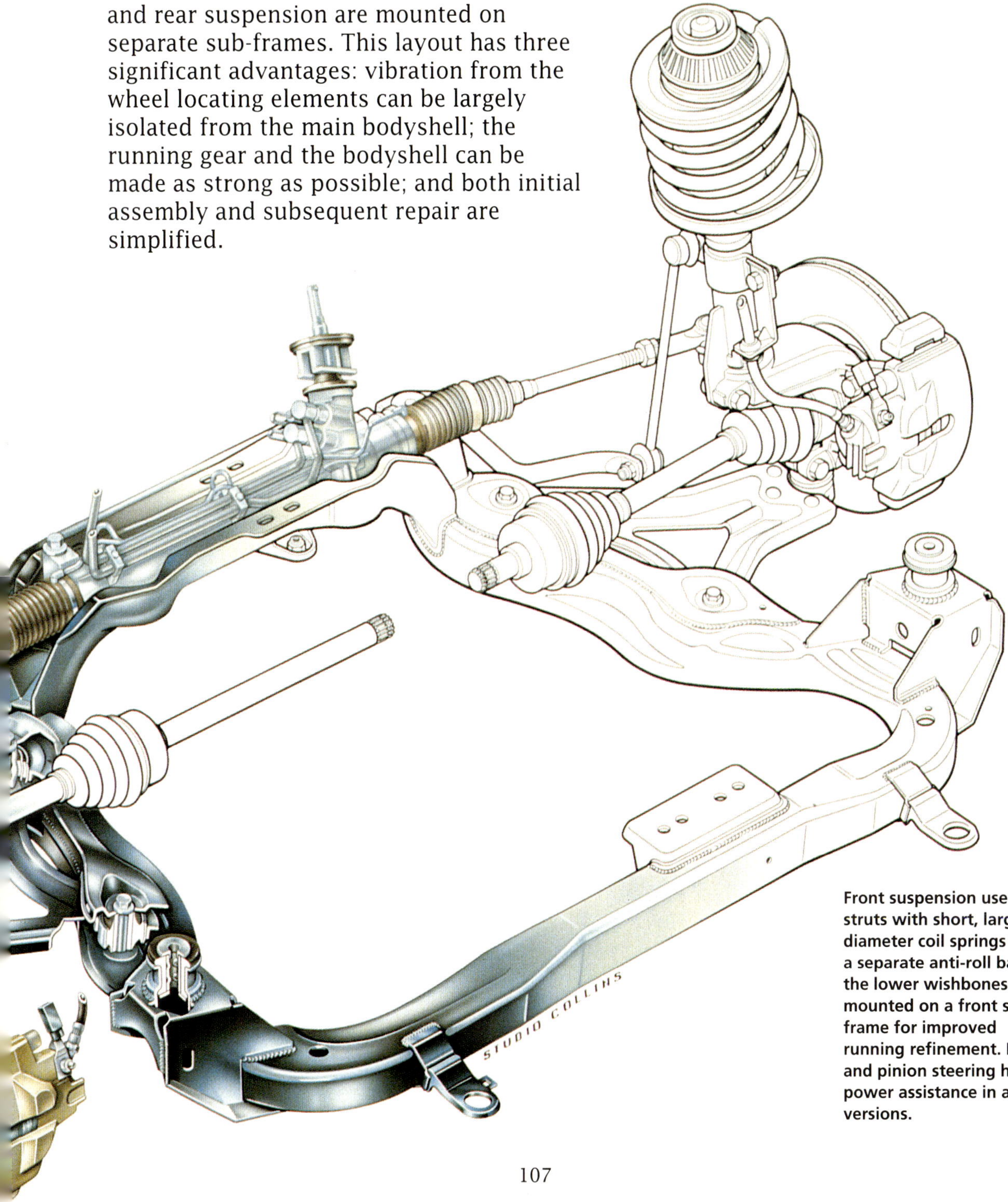

Front suspension uses struts with short, large diameter coil springs and a separate anti-roll bar; the lower wishbones are mounted on a front sub-frame for improved running refinement. Rack and pinion steering has power assistance in all versions.

CDW27's front suspension consists of triangulated lower wishbones, MacPherson spring and damper struts and an anti-roll bar. At the rear, the design problem was more complex because of the car's various body designs. Saloons and hatchbacks have MacPherson struts in combination with 'Quadralink' suspension plus transverse arms and an anti-roll bar. The four-wheel-drive version, with its rear-axle differential and drive-shafts, needs a different layout. As does the estate car, for which a low, unobstructed floor area is essential. The estate has to have, for each rear wheel, a long trailing arm, three lateral control arms and a separate spring and damper.

So there are four combinations of rear suspension - saloon/hatch or estate, two-wheel drive or four-wheel drive.

Mondeo is the first European Ford model to be offered with a variable suspension system, using computer control to maintain high standards of road behaviour regardless of load and road conditions. This Adaptive Damping System (ADS) responds within 20 milliseconds to every change in the road surface, and can be set to 'auto' or 'sport'. But even when 'auto' is selected, the system switches automatically to the stiffer 'sport' setting during hard

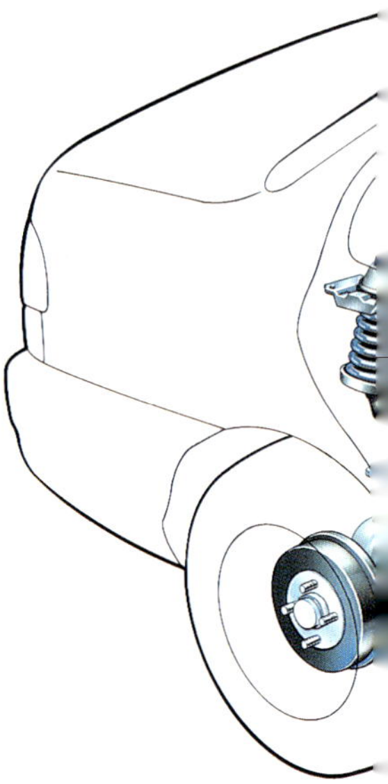

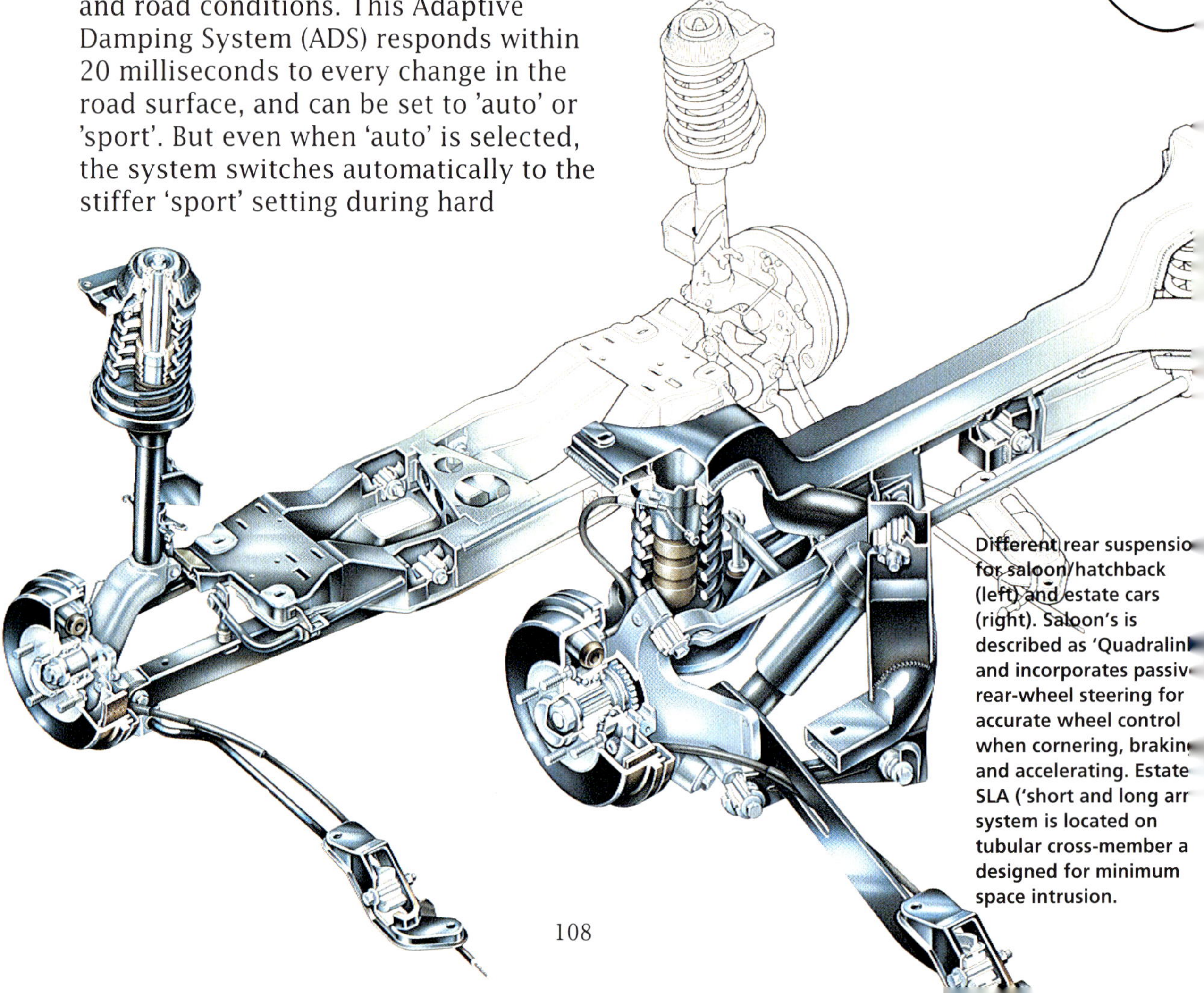

Different rear suspensio for saloon/hatchback (left) and estate cars (right). Saloon's is described as 'Quadralin and incorporates passiv rear-wheel steering for accurate wheel control when cornering, brakin and accelerating. Estate SLA ('short and long arr system is located on tubular cross-member a designed for minimum space intrusion.

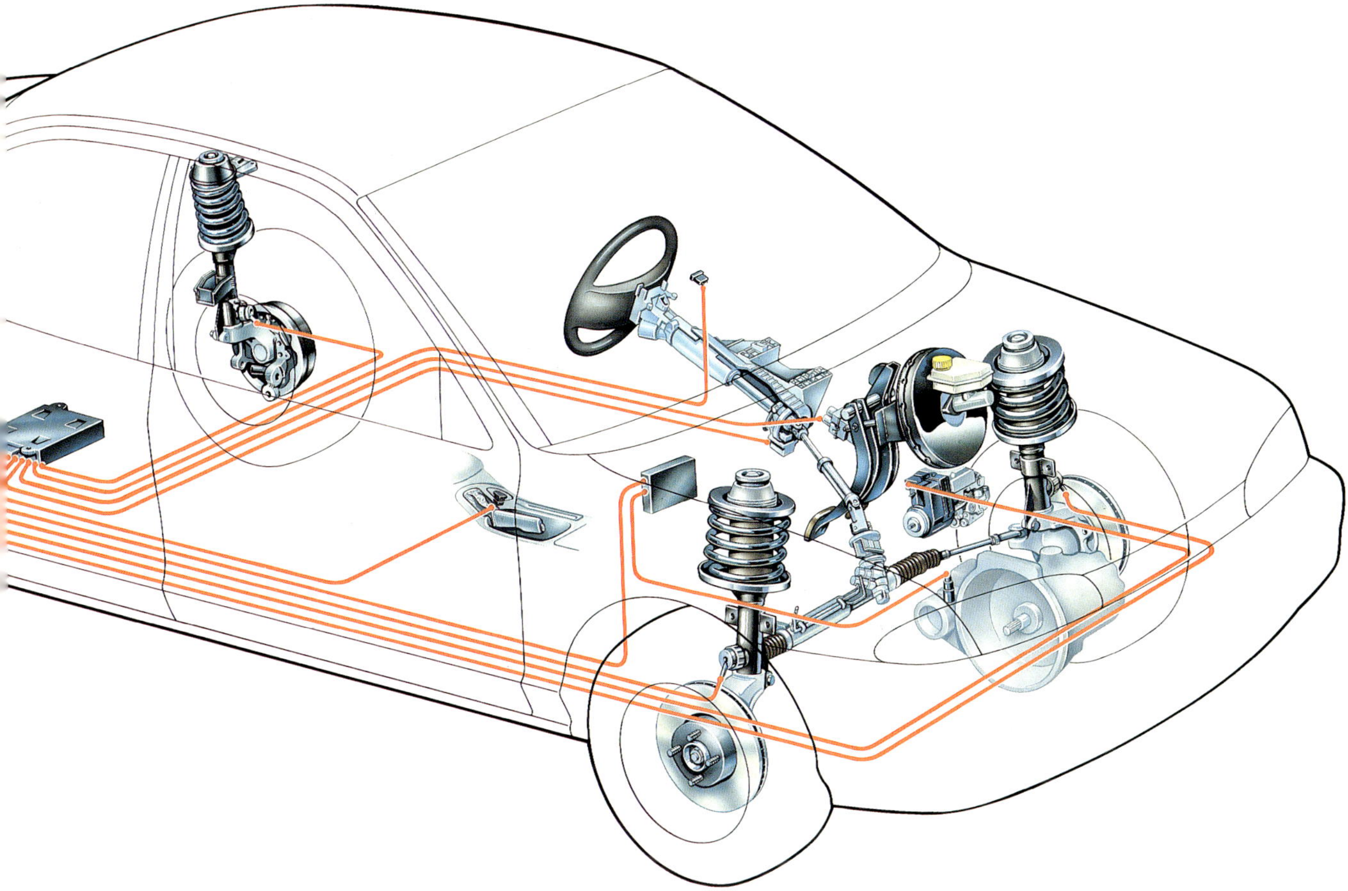

cornering, acceleration or braking, and at speeds in excess of 100 mph (160 km/h) and reverts back to the 'comfort' setting only after the manoeuvre is completed or the speed has fallen below 93 mph (150 km/h).

The car's brakes, too, are in accordance with the very latest technical standards: most models have discs all round, internally ventilated at the front and also at the rear on the four-wheel drive versions. Anti-lock braking (ABS) is available on all versions, and standard on those of higher specification.

Mondeo's ABS comes from the Bendix company. Its novel feature is the single, compact, lightweight element in the engine compartment which integrates the hydraulic and electronic components. This unit not only prevents the brakes from locking during normal and panic applications, but also helps to control the four-wheel-drive system, where specified.

Four-wheel drive Mondeos dispense with such heavy and complex assemblies as the Torsen centre differential or viscous couplings. During development work it became obvious that any

Mondeo's ABS system has integrated hydraulic and electronic control unit. The system can also direct an ATC traction control system.

wheelspin could be kept under control by the traction control system and by reducing the throttle opening and applying the brake briefly at the affected wheel. In other words, if an ABS sensor detects that its wheel is starting to spin, it activates the brake at that wheel and puts an end to the spinning movement before it can build up. The engine's power output can also be reduced to prevent wheelspin.

As we know, automobile safety has two aspects. Passive safety means the features that take effect without any action on the driver's part, whereas active safety includes those things which the driver can influence - in other words which help to avoid or minimize an accident.

ABS has been described as the greatest active safety development of the 1980s. Ford embraced it wholeheartedly - and were the first to fit it as standard equipment, on the Granada/Scorpio range. The next safety advance is in passive occupant protection: the air bag.

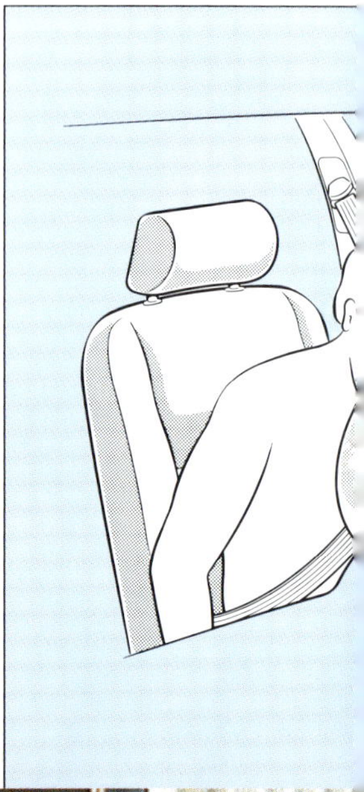

EXTRA SAFETY IN THE BAG

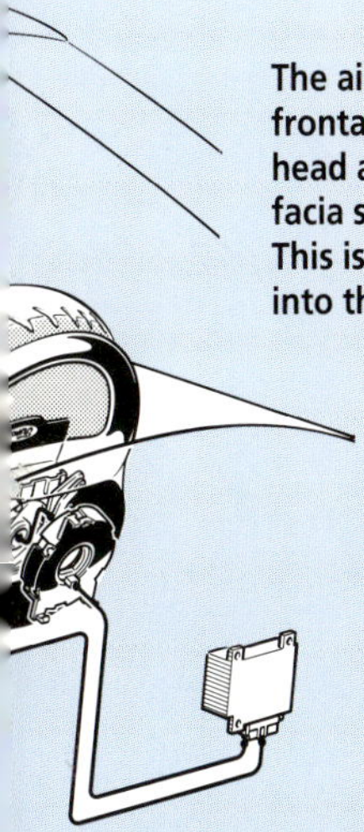

The air bag is designed to inflate instantaneously in a frontal crash and provide a cushion to guard against head and upper-torso injury from the steering wheel and facia structure.
This is a complex device. The bag itself is tightly folded into the steering wheel hub and inflated by gases from a chemical charge, prompted by electronic and mechanical inertia sensors on the vehicle body. A frontal or semi-frontal impact at over 30 km/h (18.6 mph) triggers the system. Within 40 milliseconds it swells up and then deflates progressively, cushioning a driver being thrown forward towards the steering wheel and windscreen.
In the USA, where there is not the same degree of seat belt wearing as in Europe, air bags meet the requirement for a 'passive restraint' system. There, more than 2 million Fords equipped with air bags are in service. In Europe the air bag is designed to work in conjunction with lap-and-diagonal safety belts. Ford fits a 30-litre 'Euro-bag', smaller in volume and quicker-acting than the US type. 'Grabbers', preventing slack in the seat belts, and spring-loaded pre-tensioners which tighten the belts under severe deceleration, work with the air bag to provide the best possible occupant restraint.
John Oldfield, Ford of Europe's vice president of product development, says that this combination of measures brings a substantial decrease in impact forces. Mondeo is the first all-new car developed to meet Ford's new safety strategy, a set of company standards that are considerably tougher than those required by law around the world.

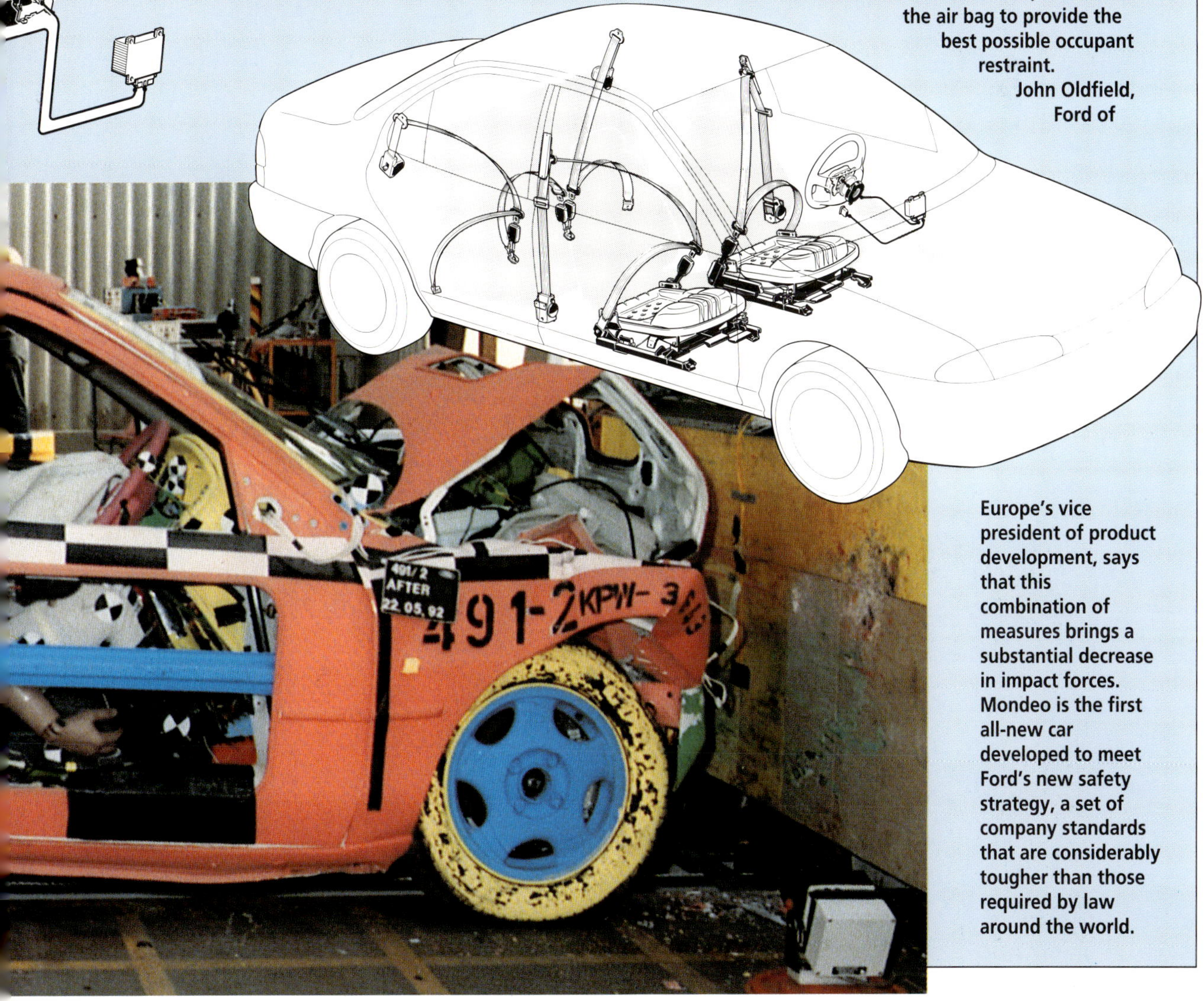

CDW27 prototypes were tested under the most severe conditions, from the searing heat of Arizona to the freezing cold of Finland.

Mondeo becomes the first car in its class to include an air bag as standard. Initially, this is on the driver's side only - where it is proved to be of the greatest benefit, protecting against chest and head injury from the steering wheel - but a passenger side air bag can be ordered as an option.

Seat belt 'grabbers' and pre-tensioners, variable-height belt pivots, 'anti-dive' seat design, door beams and side-panel reinforcements, are associated features that make up Mondeo's safety package.

But that's more than just hardware. There are many unsung heroes in a car's development. It doesn't make headlines when ergonomics experts investigate the best possible positions for the controls and instruments. Or if test drivers cover hundreds of thousands of miles in search of the best possible steering response. Safety is the effective interplay of many factors, all aimed at creating a sense of relaxed, complete control on the driver's part.

Driving a car is no longer the fundamental pleasure that it once was. Today's cars need advanced design if they are to cope with changing traffic conditions through the 1990s. With CDW27, Ford's engineers aimed for and applied all their technological resources to achieve the highest standards in every area. Their first truly global car could demand nothing less. ●

Ford

All new cars - in Europe, Japan, and the USA - must meet strict exhaust emissions regulations. But the responsibility of manufacturers in protecting the environment goes far beyond fitting their cars with catalytic convertors. Conservation of resources and recycling of waste are now vital issues.

THE GREEN OFFENSIVE

● Martin Wyler

In the industrial countries which we consider to be civilized, it is only quite recently that people have re-discovered the need to protect the environment, and come to appreciate that this policy, if correctly applied, can make our planet a far better place to live.

Over the past 10 years individual motorized transport has been the target of violent criticism in many European countries.

It was back in the 1940s, in the United States of America, that the connection was first made between air pollution and the automobile. But not until many years after the Second World War was the US Clean Air Act able to establish a firm basis for legislation restricting automobile exhaust emission levels.

Some countries in the 'Old World' waited until the second half of the last decade before following the American example. By then, in addition to the specific problem of reducing pollution from motor vehicle exhausts, there was a concern for environmental protection in a broader sense. Manufacturers became closely involved in a process that not only governed the manufacture of their products but also their eventual scrapping and recycling.

The European Community authorities have acquired the task of bringing unity into the customary tangle of regulations and policies that arises in such situations. Despite cries of protest from some manufacturers over a number of years, stringent overall exhaust emission limits for new vehicles were applied throughout the EC from 1 January 1993.

These regulations are equivalent to, but not the same as, the prevailing Federal standards in the United States - generally referred to as 'US 83'. They set maximum levels, measured in a standard test, for carbon monoxide (CO), hydrocarbons (HC) and oxides of nitrogen (NOx) in the exhaust. The standards represent a reduction of between 80 and 90 per cent of these harmful pollutants compared with an engine without emission control.

Some manufacturers struggled to present an attractive model range complying with these new regulations before the deadline. Ford of Europe is among those who have for some time, in appropriate markets, offered a range of low-pollution models with catalytic converters; in the summer of 1992 the company switched over its production generally to catalyst-equipped cars.

Mondeos will be available only in catalyst form from the outset. All petrol-engined models have a three-way catalyst system with electronic control. The diesel-engined Mondeo 1.8 TDI, already exceptionally pro-environmental, can be equipped with an oxidation catalyst to reduce further its already low exhaust emissions.

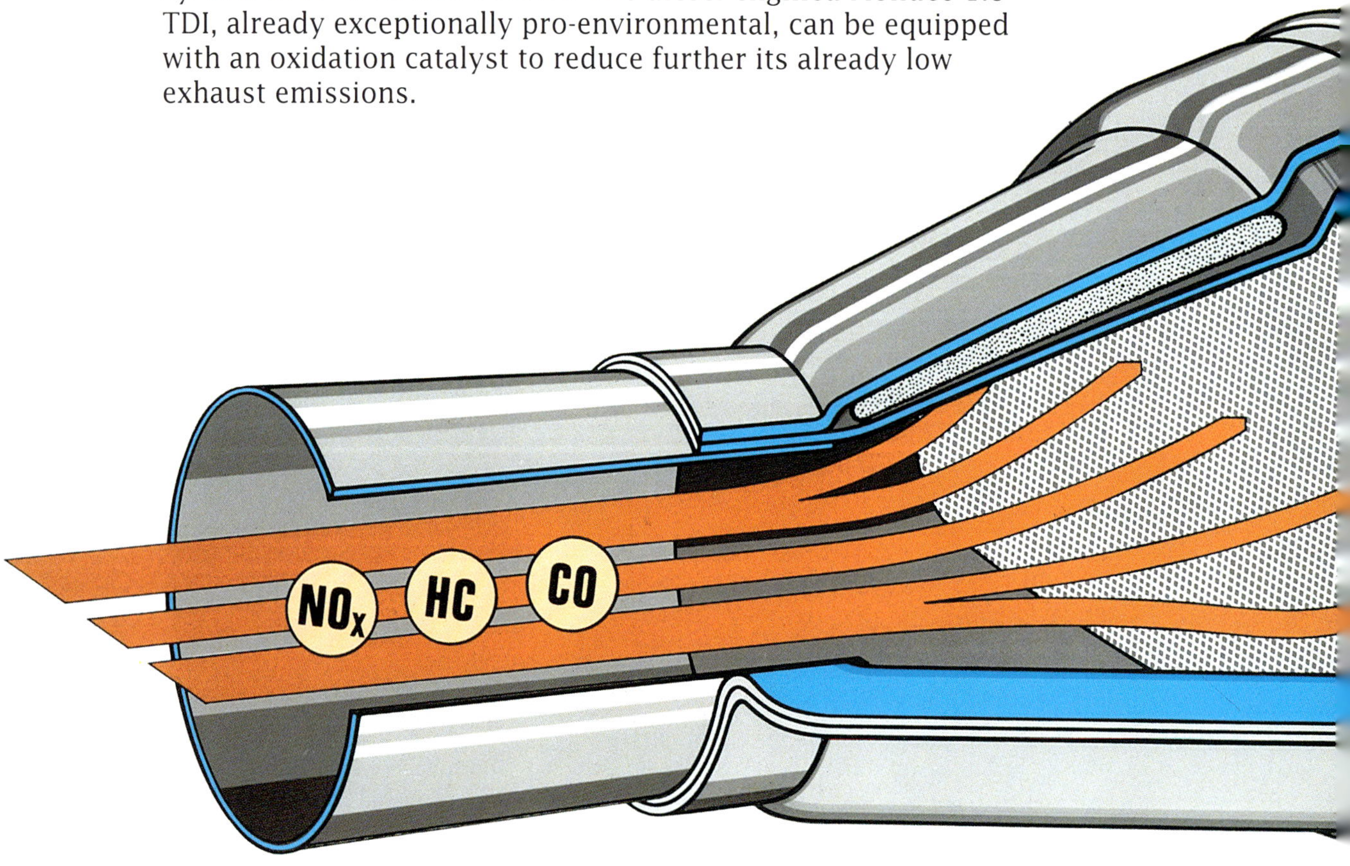

Exhaust emission control began rather hesitantly in the 1960s with closed-circuit engine breathers, but now it has developed into a complex regulating system that could never function without electronics and an on-board computer. The valves returning crankcase fumes to the air intake that were so simple in the early days, are now controlled by the same electrical and electronic circuitry that directs the engine's ignition timing and fuel injection volumes and can even influence the automatic transmission shift points. In Ford's case, this management system is run by the well-known EEC-IV control unit.

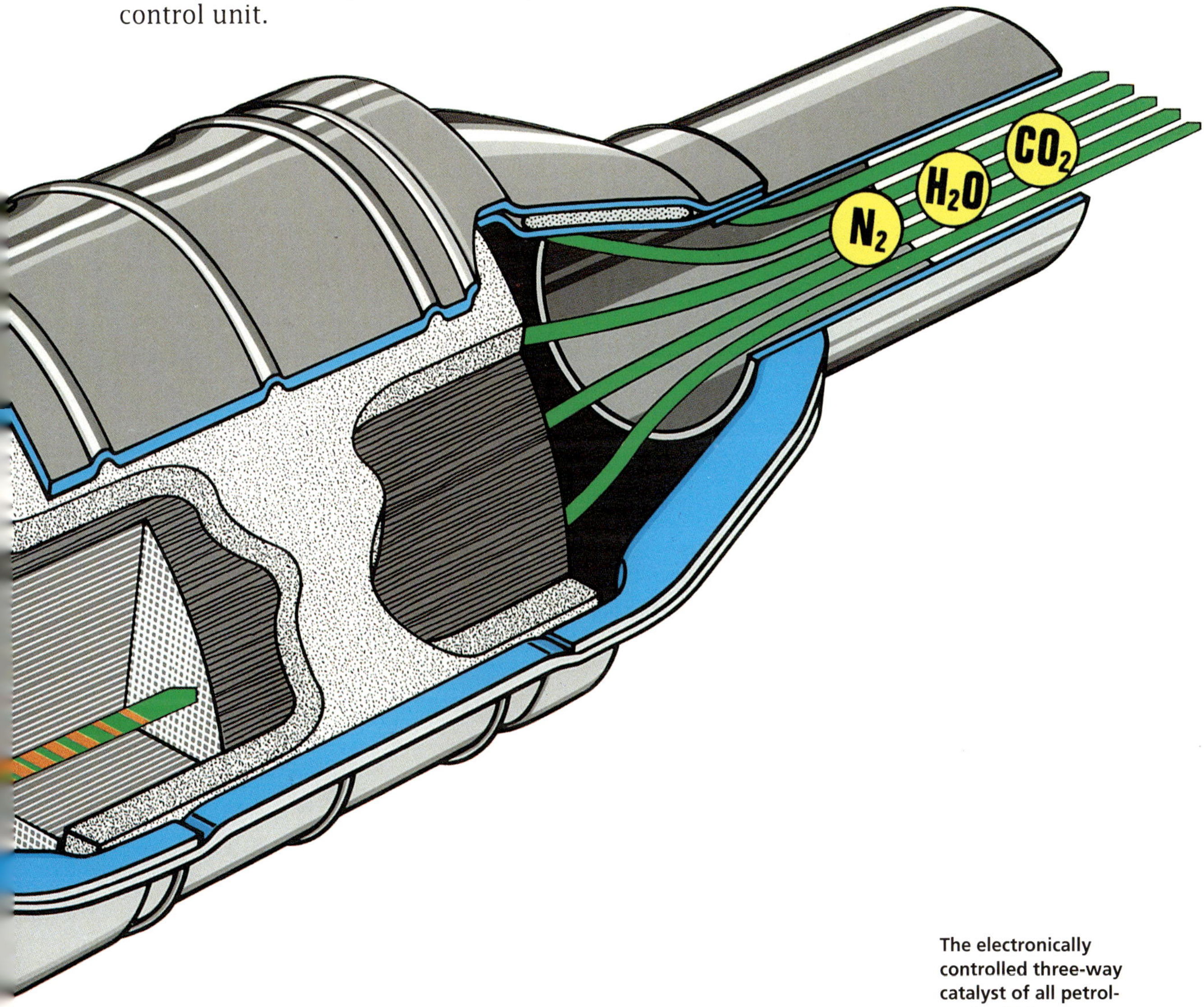

The electronically controlled three-way catalyst of all petrol-engined Mondeos.

How did such advanced emission control systems develop? Although legislation limiting exhaust pollution started in the USA, research into basic principles and many of the components - for examples, the 'Lambda probe' oxygen sensor and the closed-loop catalytic converter - took place to a large extent in Europe.

The initial idea was to satisfy the statutory pollution limits by conventional means, by moderating the ignition timing, adopting a leaner mixture (the 'lean burn' principle) or installing mechanical purification systems such as air injection into the exhaust manifold, and exhaust gas recirculation. With some or all of these measures in use, the engines lost most of their will to live. They suffered from acute power loss and, more specifically, a drop in torque that could only be compensated for by increasing the engine's displacement. That, in turn, had a negative effect on fuel consumption, which in some cases became startlingly high. Thermo-reactors were also tried, but they also incurred a severe loss of operating economy. When the so-called Energy Crisis of the mid-1970s came upon us, it did at least serve to demonstrate that we had a duty to use fossil fuel resources as economically as possible.

A new double act then came on the scene: the closed-loop catalytic converter and oxygen sensor. These devices proved capable of achieving excellent exhaust purification without a very significant increase in fuel consumption. There was, of course, a catch: unleaded fuel first had to be introduced, or else the lead would have destroyed the precious metals in the catalyst coating. That was quite clearly the greater of two financial evils, and therefore the United States of America led the way by rapidly achieving nationwide availability of unleaded gasoline.

As fuel prices rose in Europe, the authorities watched developments across the Atlantic but seemed disinclined to follow suit. Instead, efforts were concentrated on producing exceptionally economical engines and lean-burn concepts, in the hope that the problem could be solved without the unfortunate car owners having to dig even deeper into their pockets. Meanwhile, America's anti-pollution legislators had the bit between their teeth, their task being made easier by continuing low fuel prices. New fleet consumption laws (CAFE, for Corporate Average Fuel Economy) stimulated further research. In due course, new technologies made the oxygen sensor and the three-way catalytic converter feasible for smaller-engined cars as well.

In the second half of the 1980s, Europe's interest became concentrated on the 1983 US emission control legislation, and in certain European countries (Switzerland, Sweden, Germany, Austria) these severe limits were introduced for vehicle type approval in parallel with the existing ECE (European) values. Switzerland went a step further and passed its own set of exhaust emission laws from October 1982 onwards, with a further increase in severity four years later, though this move was rendered less significant by the more general adoption of the US 83 standard.

With the introduction of the EC rules in January 1993, emissions limits on both sides of the Atlantic become almost the

same. Without the oxygen probe and the closed-loop, three-way catalytic converter there would be little chance of complying with these extremely tight limits. Although the method of control to be used is not laid out in the regulations, no other technique has proved capable of yielding the same results.

In developing their project CDW27, Ford has applied the latest emission control technology to a completely new four-cylinder engine, the Zeta family. With double overhead camshafts, four valves per cylinder and sequential multipoint fuel injection, allied to the EEC-IV electronic engine management system, this design offers the best possible prospects for an effective combination of minimum pollution and moderate fuel consumption.

The closed-loop, three-way 'cat' operates in conjunction with a Lambda probe that controls the strength of the fuel-air mixture according to the amount of oxygen in the exhaust. Here it is also backed by a pulse-air system that injects air during the warm-up phase, and also by an exhaust gas recirculation (EGR) system, limited to a level of 15 per cent, which achieves a further worthwhile drop in NOx (oxides of nitrogen) emissions.

It is not only Mondeo's spark-ignition engines that represent the peak of technical progress. The 1.8 TDI model is powered by a 1.8-litre turbocharged diesel engine with charge-air intercooling (TDI stands for Turbo Diesel Intercooler). Also

Recycling - dismantling a facia at Ford's new facility in Niehl, near Cologne.

equipped with EGR and an oxidizing catalytic converter, this forced-aspiration compression-ignition engine complies with the toughest exhaust emission legislation in force anywhere in the world.

As the industrial era in which we live progresses, and as it becomes increasingly obvious that we must go easy on our remaining natural fuel resources and avoid contaminating the air we breathe, the actual production processes from which we obtain industrial and consumer products are being ever-more closely scrutinized.

It is a well-known and welcome fact that asbestos is no longer used for brake and clutch linings. Action is also being taken to banish heavy metals such as lead and cadmium from our motor vehicles, as well as the CFC group of chemicals used until now in making foam plastic seat cushions, instrument panels, and bumpers. Even the air conditioning systems of the future will use refrigerants without potentially hazardous CFCs. Water-soluble paints cut the proportion of solvents to a minimum without the any loss of quality in the paint finish.

At the Ford Motor Company's production facilities around the world, pro-environmental methods have been introduced

Plastic bumper casings may reappear as concealed bumper reinforcements.

New procedures for recycling include interior trim materials as well as metal parts from engines and car bodywork.

steadily over a period of many years. Effluent treatment plants purify millions of cubic metres of water every year and return it to the production process, mixing it with the minimum practicable quantity of fresh water. In its incineration furnaces, Ford disposes of thousands of tons of waste material and generates sufficient energy to satisfy up to a quarter of its total heating requirements.

Closed-circuit plants of this kind and energy management systems greatly reduce the amount of energy that has to be generated for the production process. They are a neat and effective way of coping with peak electric power and heating demand. Furthermore, thermal coupling reduces pollutant emissions in other areas, since, depending on where the plant is situated, less fuel (heating oil, lignite, or natural gas) will then be consumed. This results in a very significant drop in sulphur and carbon dioxide emissions (SO_2 and CO_2). The Ford plant in Saarlouis is the first European automobile production facility to operate with no input of primary energy at all.

As a follow-up to an environmentally acceptable production process and the low-pollution vehicles which it brings forth, recycling is now accepted as the logical factor to round off the product cycle.

Ever since the early days of the automobile, a degree of recycling has taken place without attracting any undue attention. The celebrated Ford Model T, for instance, more than 15 million of which made, was a splendidly simple design that could be dismantled completely without a great deal of manual effort. The materials that resulted could easily be sorted and disposed of in an appropriate way: iron and steel were returned to the foundry furnaces and wood and fabrics were burned as a welcome source of heat in winter.

As the years wore on, however, automobiles grew increasingly luxurious, with wood being used for the interior trim of only the finest and correspondingly expensive models. New materials appeared; light alloys and precious metals began to feature in automobile engineering and were soon followed by plastics and other synthetics. In modern cars these perform a great variety of tasks, but their correct disposal is something to which both scrap dealers and the automobile industry are having to devote much thought. Large quantities of old bumpers, facia panels and similar components simply cannot be stacked somewhere in the countryside for ever more: they must be disposed of efficiently or, better still, recycled.

CDW27 is one of the motor industry's first new models to be planned from the start for maximum component recycling. The average model consists of 78 per cent metals (steel, cast-iron, aluminium, copper etc.) and about 1 per cent liquids (oil and water). That leaves 21 per cent, made up of 8.2 per cent plastics, 4.9 per cent rubber (including the tyres), 2.8 per cent glass, 2 per cent sound-absorbing material and felt, 1.4 per cent foam material, 1 per cent paint and sealants, and 0.4 per cent carpets and vinyl.

Different procedures are needed to recycle or dispose of each of these categories at the end of the car's useful life. Ford has, therefore, set up a dismantling line at its plant in Niehl, a suburb of the German city of Cologne. This line has two purposes to fulfil: to supply know-how on ways of designing future models so that they are even easier to recycle, and to work out procedures for dismantling Ford models going back to 1975 as efficiently as possible, and sorting the parts according to the materials they are made of. Like the car's workshop manual, to which mechanics refer for maintenance and repair instructions, licensed recycling companies will be issued with a dismantling manual for future Ford models. A particularly important task will be to separate and re-use the various plastics, which are now being marked to identify them.

Many plastics can be returned to the production cycle, even if only for products not requiring the same standard of quality as the original: bumpers, for instance, can reappear as concealed bumper reinforcements, wheel covers are transformed into air entry grilles or storage trays, noise-insulating materials are reprocessed to yield new felt, and a modern plastic fuel tank can contribute towards the manufacture of a new tank of the latest pattern.

The 'green offensive' for which Ford has now given the starting signal is not restricted to the company itself and its products, but involves dealers and service partners too. Environmental protection, the use of alternative energy sources and, in particular, the creation of pleasant working conditions for many thousands of employees are objectives which the Ford Motor Company is pursuing in parallel with its policy of building economical, environmentally acceptable automobiles suited to modern driving conditions. ●

To achieve maximum benefit from the global programme, Ford's production experts sought to reduce the complexity of manufacturing CDW27. Just two factories will build the cars. One, Genk in Belgium, was also responsible for all the prototypes and pre-production work. It can claim to be 'The Home of Mondeo'.

PROGRESS ON THE LINE

● **Vic Heylen**

In the early 1960s, Ford of Europe, like most European car manufacturers, started to look for suitable sites to build a new plant to meet a spectacularly growing demand for cars.

When Genk, in the Belgium province of Limburg, was finally chosen, it was for a wide variety of reasons. Genk is situated close (80 miles) to Ford's German headquarters in Cologne, and near to the Cologne-Antwerp motorway, which runs further all the way to Zeebrugge, Belgium's North Sea port.

Although the region did not have an automotive history or a local source of components, Genk proved to be a very good choice. Because of the closure of collieries - coal mining had been the main industrial activity in the area - there was a large reserve of industrious workers.

On February 16, 1962 a deal was closed with the City of Genk for the purchase of 176 hectares of land and an option for a further 133 hectares for possible future expansion. As soon as construction work started, newly-recruited employees were sent to the Ford plant in Cologne to be trained in car making.

Two years after breaking ground, production started. Within five years, in September 1969, the one millionth vehicle left the Genk assembly line.

By the late 1980s, Genk was producing the Ford Sierra and the Transit van. Sierras were also made in Britain, at Dagenham, but the majority came from Genk. Ford of Europe's directors had more than once asked themselves the question: 'Why not

The Genk plant - home of Mondeo. All European versions of CDW27 are assembled here.

rationalize Sierra production by concentrating it in one plant?'

As the time for a Sierra successor approached, this became a more important issue. The decision was made: Sierra production would be concentrated in Genk and the new car, code-named CDW27, would be built there exclusively.

Apart from the low-volume Granada/Scorpio, all other Ford models are at least double-sourced. To produce a very important car like CDW27, which would represent about one fourth of Ford of Europe's total passenger car output, in only one plant was a major policy decision.

But Ford was re-examining the way it did business, seeking not only to cut costs but also to spend its money in the most effective way. Having to prepare and to tool up two plants for the production of CDW27, would certainly be more expensive than producing the car in Genk alone, even when the investment needed to increase capacity from 1,500 to 1,970 cars a day was taken into account.

Peter Heller, operations manager at Genk, recommended that the best way to achieve the greater output was to expand the plant and to introduce a third, permanent night shift. This meant a major change in working practices which needed to be discussed with the representatives of the employees. The unions not only accepted the change, they positively supported the plan. They recognized that becoming the sole European source for the new car would strengthen the strategic importance of the plant and ensure job security. Also 1,500 new jobs would be created in a region where unemployment was still high. The Genk plant had become the most important employer in the region, with some 14,000 people in the plant and a few thousand more working for suppliers nearby.

To make the proposal more attractive for the plant's employees, it was proposed to limit the night shift to 6.4 hours. In July 1990 an agreement between management and union representatives was reached.

$600 million would be needed to expand Genk and set up the production facilities for CDW27. A great deal of that investment went into the new press shop. This was not just to accommodate the extra volume but because the most modern, state-of-the-art transfer presses were required to provide the very high degree of quality demanded by the CDW27 programme. The body assembly line also had to be modernized and further automated - covering 95 per cent of all the spot-welds, compared with 85 per cent on the previous model.

One of the major headaches of any car producer is to predict which way the market will go and to plan production schedules in order to be able to meet market requirements. Peter Heller observes: 'If you have planned 70 per cent of production to be hatchbacks and the market suddenly decides that it wants mainly notchbacks (saloons), that means that you are in trouble'.

Production flexibility is the answer. The Genk plant has two body assembly lines: the Number 1 line is equipped to build both notchbacks and hatchbacks in any mix, while the Number 2

The body assembly line was modernized and further automated for Mondeo 95 percent of all spot-welds are done by robots.

INSTR.-PNL.S

Pilot production - pre-production CDW27s being assembled at a special facility within the Genk plant.

line produces notchbacks, hatchbacks and estate wagons. 'With this arrangement, if the market wanted only notchbacks or only hatchbacks, it would not worry us too much', says Heller. 'We would not even blink an eye if half of the cars sold were to be estate wagons'.

Since the Genk plant started the production of the Sierra in 1982, a lot has changed in the world of automobile production technology. Western car manufacturers find themselves facing strong Japanese competition. New buzz-words have become part of the car makers' vocabulary: Just-in-Time, Zero Stock, Zero Defect, Teamwork.

CDW27 came at the right time to take advantage of new production technology and to introduce innovative and efficient management systems.

Heller is a strong believer in the cost-reducing possibilities of the Just-in-Time (JIT) delivery system. He discussed the idea with suppliers to the Genk plant and was able to convince several of them of its advantages.

Impressive: $ 600 million was invested in the new press shop, which incorporates the most modern transfer presses currently available.

JIT means that the plant only maintains a minimum materials stock. Suppliers effectively deliver components direct to the assembly line, and at the exact moment that they are needed: just in time. Because JIT reduces inventory, it reduces costs related to holding large stocks of components and materials, and it eliminates double-handling. It reduces the need for a large and costly storage area. And it improves quality; with only a few parts in the pipeline, not many mistakes can get through...

JIT works best with suppliers located in the direct vicinity of the plant. Because it reduces transport costs and offers the advantages of shorter lines of communication between the assembly line and the components plant, JIT gives the opportunity to react quickly to quality and management concerns.

By rearranging the traffic flow around the plant and the parking area, Heller managed to free a large piece of otherwise unused land which was sold back to the city of Genk on condition that they would only sell it to Ford's suppliers who were willing to invest there. Several have already taken up the opportunity.

The Genk management are also great believers in teamwork. 'Working as a team on an assembly line, instead of as individuals', explains personnel manager Marcel Kellen, 'increases the worker's sense of responsibility, it adds substance

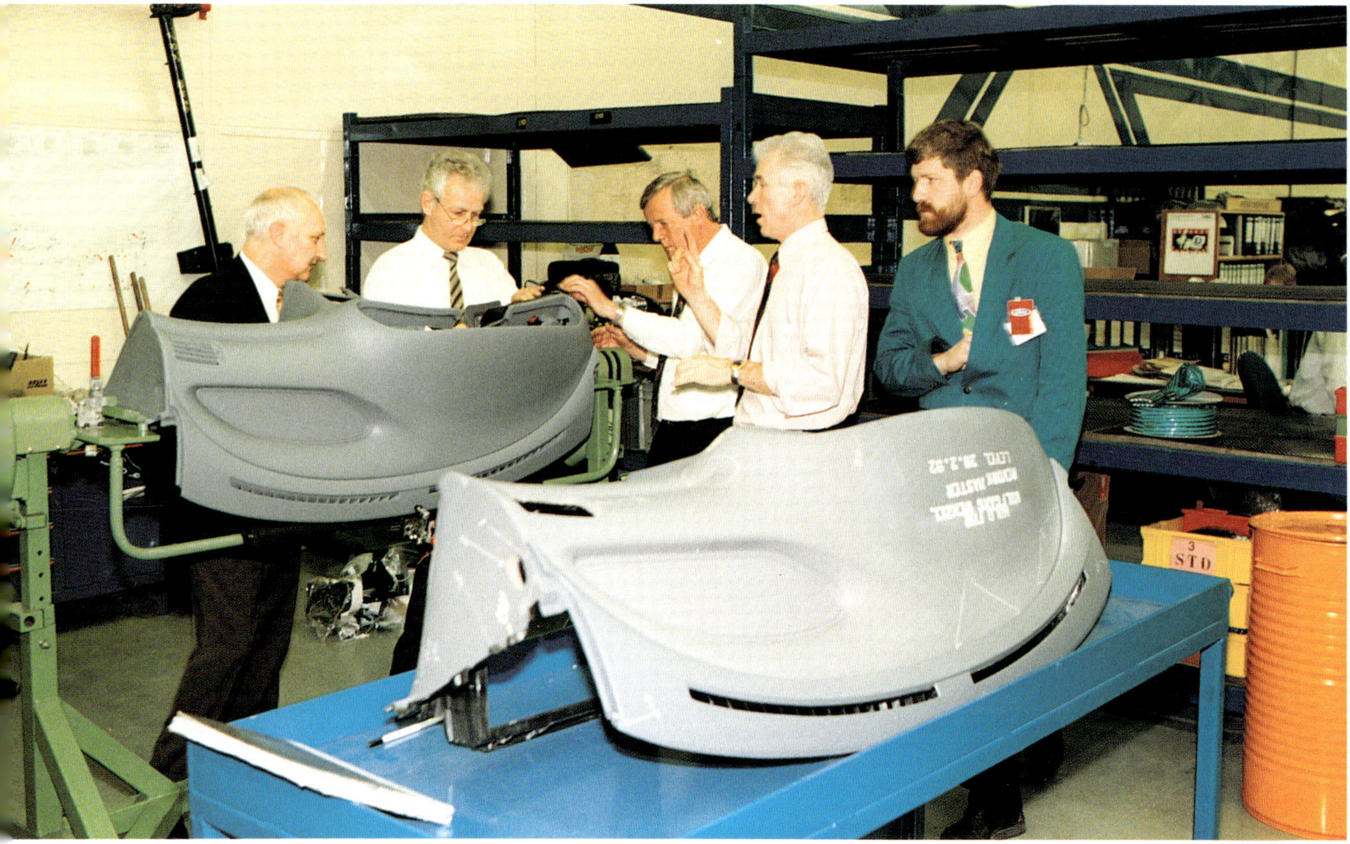

to his work and enhances job satisfaction. The end result is a job well done and consequently a high quality product. We certainly believe that customer satisfaction, which is our final goal, starts with workers' satisfaction.'

Teamwork does not replace the classic assembly line. It is a system where a team assumes total responsibility for a number of assembly line jobs, each of which, in the old system, were the responsibility of a single worker. Teamwork is not a major change in production technology; it is, however, a different human and organisational approach to assembly line work.

Marcel Kellen is convinced that gradual introduction of teamwork is essential to achieve a level of productivity and quality which is needed to face the Japanese competition. 'We already have teamwork in our press shop, operating through the weekend. It works wonderfully well. I've never seen such motivation and involvement as in these teams.'

Teamwork is not something that can be introduced overnight in the whole plant. 'It has to grow from the shop floor upwards', says Peter Heller, 'And it is'.

Delegating more responsibility to the workers also means that the decision making process has to move downward, away from the top and toward the shop floor. Peter Heller believes that it is all a matter of trust: 'Why shouldn't we give a teamworker or a teamleader the responsibility to stop the assembly line when he is being confronted with a problem that he cannot solve on his own? A problem which could affect the quality of the product, if it is not taken care of, there and then? I am in favour of installing emergency pull cords running above the assembly line, which workers can pull to stop the line, if needed. Or start a warning light to flash for help in case of a minor problem.'

Genk has been living with CDW27 from its earliest days. Not only was the plant to be prepared to build the cars but all the pre-production work was done there. The normal procedure within Ford is that a specialized pilot plant - at Niehl, near Cologne, or Aveley in southern England - builds the prototypes, evaluates them, makes the necessary changes and corrections, and delivers the definitive car to the production plant.

For CDW27, the prototypes were built at Genk. This had the advantage that the people who would be responsibile for manufacturing the final car were involved in guiding it from drawing board (or, rather, computer screen) to the assembly line.

Launch team at work. By basing all pre-production activities at Genk, problems could be resolved more quickly than for previous projects.

Masterminding the CDW27 launch team at Genk were co-ordinator Hans Schardt, Malcolm Thomas from product engineering, and Chris Boyce from the programme office.

Hans Schardt remembers his first contact with the new project. 'When I came back from holiday in August 1991, I had a message to contact Marc Noppe, first thing next morning'. Belgian-born Noppe was at that time director of body assembly operations in Europe. 'I really had no idea what it was all about and why there was such urgency. But the next morning I found out.'

Noppe told Schardt that he wanted him to take responsibility for the manufacturing side of the CDW27 launch operations, bringing together all the production areas of the Genk plant. This was to include building some 250 prototypes.

Until that time, Genk had been simply a production plant. It had no prototype shop, nor anyone with experience of building prototypes. There were no development engineers on site.

So a new team of engineers and specialists in the different disciplines had to be put together. One thing in their favour was that Ford's Lommel proving grounds, where the prototypes would be tested, was only 25 miles away.

Time was the enemy. Job 1, the first production car off the assembly line, was scheduled for the end of November, 1992 - just 16 months away.

The launch team was made up of 120 engineers, recruited from Dunton and Cologne. Some would come to live near Genk for the duration, others found temporary accommodation during the week and went home to Germany or England at the weekend. Ford company flights, running daily from Stansted in Essex to Cologne, had to be rescheduled to include a more frequent service to Maastricht, the nearest airport to Genk. Those involved in the development of Ford's global car became familiar with Maastricht well before it was the name on the lips of every European politician...

There were not enough offices at Genk for the launch team and their voluminous computer terminals and not enough time to build any. The solution was pre-fabricated buildings. A plot of land was prepared, and six weeks later, 1,230 square metres of offices were ready for occupation.

On 16 September 1991, the computer screens lit up, and the engineering team in Genk was in business, able to talk and swap pictures and data with their counterparts in other Ford engineering centres around the world.

CAD/CAM systems (Computer Aided Design and Computer Aided Manufacturing) are now vital to the process of bringing a new car to production, and have allowed development time to be cut considerably.

From the start it was obvious that, to get to Job 1 on time and achieve the required quality, a very tight and thorough system of reporting would have to be put together to deal with the problems and difficulties which would inevitably arise.

Problems would come when they started to build up the first cars: a socket that does not fit, a cable missing, or too short or too long, a component that is hard to assemble. Test drivers would send in reports of their findings when they put the prototypes through their paces on the Lommel test track.

'We set ourselves as an objective that 50 per cent of all problems that came up would have to be solved within 48 hours,' explains Hans Schardt. 'Ninety per cent of the remaining 50 had to be solved within 72 hours. The last 10 per cent were

THE ULTIMATE TEST

Job 1, 23 November 1992: Mondeo production has begun. But sales do not start until March 1993. In the meantime, 115 cars are already in the hands of customers, and subject to intensive use and thorough scrutiny.

It is accepted that even the most comprehensive test and development programme cannot always reproduce experiences in normal use by a wide spectrum of buyers. The customer evaluation programme, which Ford conducts with all new models in the months immediately before launch, is intended to identify and correct any unforeseen snags.

For Mondeo, this involves high-mileage drivers in Britain, Germany and Finland. They are testing a selection of four-door and five-door models, with 1.6, 1.8 and 2 litre engines and manual transmission.

Participating customers include company fleets and police forces. The programme will continue for 18 months, with a random selection of the cars being recalled for examination after three months, and all the cars being taken back by Ford after a year and replaced by identical models.

The same customers will use the replacement cars for six months, and be asked to assess any modifications made in the first phase of production.

major problems that could not be solved by the launch team and had to be sent to the technical centres at Dunton or Merkenich. We then expected the engineers there to come up with an action plan within five days.'

Every day, at 10 am, Hans Schardt, Malcolm Thomas and Chris Boyce presided over a meeting with their engineers to discuss the issues of the day and evaluate the way previous problems have been solved.

'Office mechanization is a very important tool in this project', explains Hans Schardt, 'before the meeting, everybody concerned has a detailed print-out of the problems at hand and a timing schedule for the solution to problems discussed at previous meetings.'

In the past, resolving problems during the pre-production period was often delayed by lengthy discussions which had to be taken to higher authority; things like who would pay the extra costs involved in making a late modification or rectifying a faulty component. For CDW27, such decisions were made by the launch team, which included purchasing people so that they could negotiate directly with suppliers.

Start-up production includes three variants of Mondeo - saloon, hatchback and this elegant estate car.

When they started prototype-build, the launch team invited production workers to come and have a look at the jobs they would have to perform later on the assembly line. That resulted in many useful suggestions which would make final assembly easier and therefore more efficient - and improve quality.

This points to a major advantage of having the launch team in the plant where the cars were to be built. Genk's employees saw CDW27, Mondeo, virtually grow in front of their own eyes. It gives them a sense of involvement in the whole process, creating a lot of enthusiasm for the new car.

For the launch team, only part of their job was over as the first Mondeos rolled off the Genk production line. Work continues on the US versions of CDW27, which were always planned for launch after sales had started in Europe. Several of the Genk team moved to America, to Kansas City, to assist in the pre-production phase there.

Some, however, will stay at the birthplace through Mondeo's lifetime, working with the Genk production team in its future developments and helping to maintain the high standards that they set for themselves when they embarked on project CDW27. ●

Mondeo
Mondeo

On the previous pages Ford's top of the line Mondeo, the luxury 5-doors Ghia version, fitted with special alloy wheels.

CHAPTER 8

MONDEO'S PLACE IN THE WORLD

● **Edouard Seidler**

The goal of project CDW27 is a yearly production of 650,000 cars, around 10 per cent of Ford's total output worldwide.

The American version will start to roll off the Kansas City assembly lines 15 months after the European launch. Ford expects to make 250,000 a year in the USA.

Most of the US production should result in incremental sales for Ford, because the new car will not replace any existing model. The original idea was that CDW27 - the names for America are not yet decided - would take over from the Ford Tempo and Mercury Topaz. It soon became obvious, however, that the American version of CDW27 would be further up-market than Tempo and Topaz. So these models will continue and the CDW27 variants will become additional models in Ford's range.

Ford's North American operations will further benefit by supplying some major components for the European Mondeo - notably the 2.5 litre V6 engines, expected for 1994, and automatic transmissions, which are available at the outset.

In Europe, the Mondeo takes over from the Sierra. It is produced in one factory only, Genk in Belgium. Its four-cylinder engines are made in the United Kingdom and in Germany, the manual transmissions provided by the Ford carousel, and it uses components from countries all over Europe.

Production of Sierra, launched over 10 years ago, was less than 300,000 units in 1992. Ford hopes that the Mondeo will sell up to 400,000 a year if European markets pick up and the car is highly successful.

Lindsey Halstead, who was chairman of Ford of Europe through the development of CDW27 and retires just as Mondeo is prepared for its sales launch, says: 'We kept selling the Sierra too long in its original version. Mondeo will be renewed faster and it will benefit from numerous innovations throughout its life cycle.'

Mondeo will make its public debut at the Geneva Motor Show in March 1993. It will be launched there with its various European body styles: four-door saloon, five-door hatchback, and estate car. It will be available initially with four-cylinder Zeta engines and the choice of manual or automatic transmission. Power steering is standard. So is a driver's side air-bag. Anti-lock braking will be available on every version. Not long after the launch, the turbo-diesel will join the range, and the 2.5 litre V6 petrol engine will be added to the Mondeo range 15 months later.

Jacques Nasser, who succeeded Lindsey Halstead as Ford of Europe Chairman on 1 January 1993, now leads the Ford of Europe team to launch the Mondeo to the market. He is looking forward to the challenge: 'Mondeo's elegant design, technology and high quality provides outstanding value to the customer. We believe it represents a new world-class benchmark in this highly competitive class.'

● **Arturo de Andrés.** Born in 1940. Started a Doctorate in Technical Engineering, but entered the weekly "Velocidad" as Test and Technical editor in 1962.
In 1969 he went to the weekly "Autopista" in the same function.
In 1985, still in the Luike-Motorpress Group, he changed to the monthly "Automóvil".
At present, he is technical editor of the monthly magazine "Automécanica".

● **Fulvio Cinti.** Born in Tuscany, has been a professional journalist since the beginning of the 1950s. Entered the motoring world as a sports writer for "Gazzetta del Popolo" in Turin.
In 1963, he started the first Italian magazine dedicated to automotive style: "Style Auto". In the capacity of chief editor of Turin's "Stampa Sera", he maintained his motoring interests.
In 1979 he brought into reality the idea of "Auto & Design", of which he is editor-in-chief and publisher.

● **Vic Heylen** is an Antwerp based, full time freelance writer, specialised in technical, economic and industrial aspects of the automotive industry.
Before starting a writing career, he spent most of his life in the automotive industry, mainly as an independent consultant, specialising in the development of race and rally cars.

● **Ray Hutton** celebrates 25 years in motoring journalism in 1993. His career has included racing and rallying and 10 years as editor of the weekly magazine "Autocar". Now a freelance writer and editor, based in London, he is the motoring correspondent of "The Sunday Telegraph" and European editor of the US magazine "Car and Driver".

● **Jürgen Lewandowski,** born in 1950, having studied German and English, has been senior editor of the motoring section "Auto & Verkehr" with the "Süddeutsche Zeitung" in Munich since 1979. Winner of the "Christophorus" award, he has written 26 books on motoring history and environment, as well as the development of new cars.

● **Edouard Seidler,** is a graduate of the School of Political Science (Paris) and has an MBA from UCLA (Los Angeles). Former publisher of France's daily "l'Equipe" he is now an automotive freelance journalist for 19 publications in Europe, America and Japan.
Author of 14 books. A specialist of interviews with car makers' top managers. Knight of the French Legion of Honour, Officer of the National Order of Merit. President of the European "Car of the Year" jury.

● **Martin Wyler,** born in 1943, grew up near Bern, facing the snow-mantled Alps. He studied Constructional Engineering, graduating in 1967.
Since 1968 he has worked for the well-known Swiss magazine "Automobil Revue" at the motoring section "Test & Technique", as well as for "Automobil Revue's" annual catalogue.